practical oscilloscope handbook

SECOND EDITION

HOWARD BIERMAN
PAUL BIERMAN
RUFUS TURNER

HAYDEN BOOK COMPANY, INC.
Rochelle Park, New Jersey

Library of Congress Cataloging in Publication Data

Bierman, Howard.
 Practical oscilloscope handbook.

 Includes index.
 1. Cathode ray oscilloscope—Handbooks, manuals, etc.
I. Bierman, Paul. II. Turner, Rufus P. III. Title.
TK7878.7.B53 1981   621.3815'48   81-2924
ISBN 0-8104-0851-1   AACR2

Copyright © 1964, 1981 by HAYDEN BOOK COMPANY, INC. All rights reserved. No part of this book may be reprinted, or reproduced, or utilized in any form or by any electronic, mechanical, or other means, now known or hereafter invented, including photocopying and recording, or in any information storage and retrieval system, without permission in writing from the Publisher.

Printed in the United States of America

```
        1  2  3  4  5  6  7  8  9   PRINTING
       81 82 83 84 85 86 87 88 89   YEAR
```

preface

The more complicated electronic circuits and systems become, the more valuable the oscilloscope becomes in the laboratory, production line, and service bench.

In the "early" days of electronics, when radio and audio amplifiers were the major electronic products, the oscilloscope had its place beside the multimeter, signal generator, and tube tester. But with monochrome and then color TV, digital circuits, and more sophisticated electronic products, ac and dc voltage readings had to be replaced by waveform analysis, a job for the scope. A skilled technician with a proper scope can perform tests in minutes that might take hours (and conceivably not produce results) with a multimeter.

The *Practical Oscilloscope Handbook*, Second Edition, is written for the engineer, technician, or serious hobbyist as a practical guide to set up, display, and interpret waveforms. A basic understanding of electronic fundamentals is assumed and very little theory is included. Wherever possible, clear, step-by-step instructions are presented to simplify the measurement described.

Thanks to Hewlett-Packard, Electronic Instrument Co., Inc. (EICO), Leader Instrument, and Tektronix for their assistance and photos. A particular thanks to B&K Precision, Dynascan Corp. for their practical dual-trace application information and photos.

For those who have not yet purchased a scope, or those who are ready to upgrade their present model, the last chapter on factors influencing buying decisions should be helpful.

One final note. Today's scopes offer more features and provide more functions than earlier models, and can operate almost automatically. But they still require a competent operator who knows what the equipment can do, who can set the controls properly, and who can interpret the waveform he sees.

HOWARD BIERMAN,
PAUL BIERMAN,
and
RUFUS TURNER

contents

1 First Principles of Oscilloscopes   1

    1.1 What the Oscilloscope Is **1.2** The Electron Beam
    1.3 How the Beam Is Deflected **1.4** CRT Features **1.5** Viewing
    Screens **1.6** Function of Amplifiers **1.7** Function of Sweep
    Generator **1.8** Synchronization **1.9** Blanking **1.10** Intensity
    Modulation **1.11** Oscilloscope Basic Layout

2 Oscilloscope Controls and Adjustments   33

    2.1 Delayed Time Base **2.2** How to Set Up a Single-Trace
    Oscilloscope **2.3** How to Set Up a Dual-Trace Oscilloscope
    2.4 Single-Trace Waveform Observation on a Dual-Trace Scope
    2.5 Calibrated Voltage Measurements on a Dual-Trace Scope
    2.6 Operating Precautions

3 Oscilloscope Accessories   54

    3.1 Why Special Probes Are Needed **3.2** Low-Capacitance
    Probe **3.3** Resistor-Type Voltage-Divider Probe
    3.4 Capacitor-Type Voltage-Divider Probe **3.5** Demodulator
    3.8 Ac Probe **3.6** RF Probe **3.7** Dc Voltage Calibrator
    Voltage Calibrator **3.9** Frequency (Time) Calibrator
    3.10 Electronic Switch **3.11** External Amplifiers **3.12** Trace
    Photography

4 Voltage and Current Measurement   69

    4.1 How to Voltage-Calibrate the Screen **4.2** Direct
    Measurement of Voltage **4.3** Voltage Measurement with
    Voltage Calibrator **4.4** Measuring Pulsating Voltage
    4.5 Measuring Fluctuating (Composite) Voltage **4.6** Power
    Supply Ripple **4.7** Measuring Ac and Dc Current
    4.8 Measuring Fluctuating (Composite) Current **4.9** Current
    by Probe Method **4.10** Comparing Two Waveforms on a
    Dual-Trace Scope **4.11** Differential Voltage Measurement
    Using Dual-Trace Scope

5 Frequency and Phase Measurement and Comparison 88

5.1 Use of Lissajous Figures 5.2 Use of Modulated-Ring Pattern 5.3 Use of Broken-Ring Pattern 5.4 Use of Broken-Line Pattern 5.5 Use of Sawtooth Internal Sweep 5.6 Calibrated Time Measurements 5.7 Using a Dual-Trace Scope for Frequency Divider Analysis 5.8 Divide-by-Eight Circuit Waveforms with a Dual-Trace Scope 5.9 Propagation Time Measurement Using a Dual-Trace Scope 5.10 Digital Circuit Time Relationships with a Dual-Trace Scope 5.11 Use of Lissajous Figures for Phase Measurements 5.12 Checking Inherent Phase Shift of Oscilloscope 5.13 Use of Dual Pattern 5.14 Checking Phase Angle between Current and Voltage 5.15 Checking Phase Angle between Two Currents

6 Audio Amplifier, Receiver, and Transmitter Tests and Measurements 107

6.1 Checking Wave Shape 6.2 Checking Voltage Gain or Loss 6.3 Checking Frequency Response 6.4 Checking Hum and Noise Level 6.5 Measuring Power Output 6.6 Checking Amplifier Phase Shift 6.7 Checking Distortion 6.8 Checking Intermodulation 6.9 Square-Wave Testing 6.10 Oscilloscope as AF Signal Tracer 6.11 Oscilloscope as Bridge Null Detector 6.12 Distortion Measurement with a Dual-Trace Scope 6.13 Checking Gated Ringing Circuit with Dual-Trace Scope 6.14 Delay Line Tests with a Dual-Trace Scope 6.15 Stereo Amplifier Servicing with a Dual-Trace Scope 6.16 Using a Dual-Trace Scope to Improve the Ratio of Desired-to-Undesired Signals 6.17 Amplifier Phase Shift Measurements with Dual-Trace Scope 6.18 Checking VITS (Vertical Interval Test Signal) with a Dual-Trace Scope 6.19 Visual Alignment of AM IF Amplifier 6.20 Visual Alignment of FM Detector 6.21 Visual Alignment of FM IF Amplifier 6.22 Visual Alignment of TV IF Amplifier 6.23 Visual Alignment of TV Sound IF Amplifier and Detector 6.24 Checking the Video Amplifier with Square Waves 6.25 Visual Alignment of TV Front End 6.26 Checking TV Operating Waveforms 6.27 Checking Amplitude Modulation by Sine-Wave Method 6.28 Checking Amplitude Modulation with Trapezoidal Patterns 6.29 Checking Modulator Channel 6.30 Checking Frequency Multiplier

7 Servicing the Oscilloscope 145

7.1 Keep the Instruction Manual Handy 7.2 Contact the Manufacturer 7.3 Troubleshooting Tips 7.4 Replacing a Defective Component 7.5 Calibration Hints 7.6 Troubleshooting Chart 7.7 Internal Adjustments 7.8 Checking Input Resistance and Capacitance 7.9 Checking Input Attenuator Compensation 7.10 Using a Square Wave to Check Compensation 7.11 Checking the Low-Capacitance Probe 7.12 Checking Crosstalk between V and H Amplifiers 7.13 Checking Phase Shift between V and H Amplifiers 7.14 Adjusting Voltage Calibration 7.15 Checking Sweep Linearity 7.16 Checking V and H Sweep Settings 7.17 Checking Astigmation Adjustment

8 Selecting the Oscilloscope 159

8.1 How Much Can You Afford? 8.2 Conventional Scopes versus Plug-in 8.3 Factors Affecting Buying Decision 8.4 The CRT 8.5 Scope Bandwidth 8.6 Sweep Range 8.7 Deflection Sensitivity 8.8 Time Base 8.9 Dual-Trace Scopes 8.10 The Final Decision 8.11 Renting and Leasing Options

Index 180

safety notice

Every oscilloscope contains high voltages which can electrocute you. Be extraordinarily careful when you work inside this instrument. Never work inside a live oscilloscope or operate it with its cover removed unless you are thoroughly familiar with the instrument circuit and take care to avoid touching any high-voltage point. The oscilloscope is dangerously deceptive; it has high voltage in many unexpected places.

In most oscilloscopes, as in other test instruments, the metal outer case is connected internally to the GROUND or COMMON input terminal. When this terminal is connected to a high-voltage point in a circuit under test, the entire case becomes dangerous to touch. Be alert when you *must* use the instrument in this manner.

Follow every electrical safety rule (keep one hand in your pocket, insulate yourself from ground, use insulated tools and/or rubber gloves, know the circuit you test, connect the instrument before you apply test-circuit voltage, etc.).

The cathode-ray tube is highly evacuated. If broken, it can scatter glass fragments at high speed, possibly inflicting grievous wounds. Handle this tube with care and avoid striking, scratching, or dropping it.

1

first principles of oscilloscopes

Of all electronic test instruments, the oscilloscope comes closest to being indispensable. So versatile is this device that if a technician were limited to a single instrument, his choice of an oscilloscope would, in the majority of cases, be the wisest. The oscilloscope can show a great many things about the behavior of circuits and the nature of currents and voltages in them.

1.1 What the Oscilloscope Is

Basically the oscilloscope is an electron-beam voltmeter. Its indications are produced by applied signal voltages that deflect a thin beam of electrons instead of a pointer. Since it has no significant weight, the electron beam can move several million times faster than the lightest pointer. Completely free of mechanical parts, the oscilloscope is a true electronic instrument.

In its performance, however, the oscilloscope is unlike any other voltmeter. The electron beam faithfully follows rapid variations in signal voltage and traces a visible path on a screen. In this way, rapid alternations, pulsations, or transients are reproduced and the operator can see their waveform as well as measure their amplitude. Because of its completely electronic nature, the oscilloscope can reproduce high-frequency waves which are much too fast for such electromechanical devices as direct-writing recorders, and oscillographs.

Before the oscilloscope was developed, a technician could not "see" a high-frequency waveform but had to piece together a graph of it from data laboriously taken, point by point. And even then, phase relations and distortion (immediately evident on an oscilloscope screen) had to be painstakingly calculated. But the oscilloscope has removed the "uncertainty of the unseen" and has simplified many tests and measurements. In doing so, it has become an immensely useful tool in the electronics laboratory, repair shop, radio station, and classroom. This instrument is also invaluable in any field where a phenomenon may be converted into a proportional voltage for observation—meteorology, medicine, biology, chemistry, mathematics, psychology, etc.

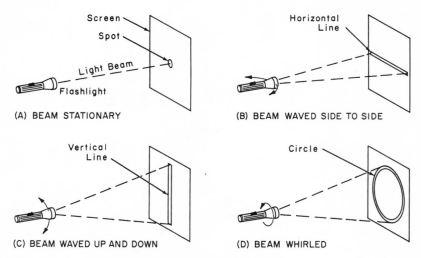

Fig. 1-1. The light-beam analogy.

The oscilloscope, then, is a kind of voltmeter which uses an electron beam instead of a pointer, and a kind of recorder which uses an electron beam instead of a pen. It saves test time by directly displaying a phenomenon.

1.2 The Electron Beam

The action of a beam of light illustrates how the electron beam works in an oscilloscope.

Point a sharply focused flashlight at a screen or wall, as in Fig. 1-1A, and the light beam will make a bright dot where it strikes the screen. Hold the flashlight still and the dot remains stationary; move it and the dot is displaced on the screen. If the movement is slow, the eye can easily follow the spot. But if the movement, always along the same line, is too fast for the eye to follow, as happens when the flashlight is waved rapidly, *persistence of vision* causes the eye to see the pattern traced by the spot. Thus, wave the flashlight from side to side to trace a horizontal line (Fig. 1-1B) and up and down for a vertical line (Fig. 1-1C), or whirl it for a circle (Fig. 1-1D). If the hand were steady and fast enough, the light beam could be used in this way for any kind of writing or drawing.

A similar action takes place in the cathode-ray tube (CRT) of an oscilloscope. The flashlight is replaced by an *electron gun*, the light beam by a narrow *electron beam*, and the external screen by the flat end of the glass tube, which is chemically coated to form a fluorescent *screen*. Figure 1-2 shows this arrangement. Here, the electron gun generates the beam which moves down the tube and strikes the screen. The screen glows (fluoresces) at the point of collision, producing a bright dot. When the

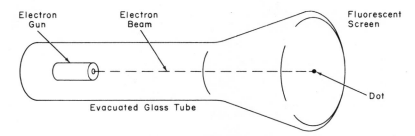

Fig. 1-2. The basic structure of the cathode-ray tube.

beam is deflected, by means of an electric or magnetic field, the dot will move accordingly to trace out a pattern.

The electron gun contains several parts arranged in this sequence:

1. A heated *cathode* out of which the beam electrons are boiled.
2. A *control electrode,* powered by a negative dc voltage, which regulates brightness of the trace.
3. A first *accelerating electrode,* powered by a positive dc voltage, which speeds up the beam.
4. A *focusing electrode,* powered by a positive dc voltage, which narrows the beam to give a thin, sharp trace.
5. A second *accelerating electrode,* powered by a positive dc voltage, which speeds up the beam.

After leaving the cathode, electrons pass through a tiny hole in each of the electrodes before they reach the screen. As you can see, the electron gun is both the source of the electron beam and an electronic system that controls the beam's focus and brightness.

1.3 How the Beam Is Deflected

The electron beam may be deflected transversely by means of an electric field *(electrostatic deflection)* or a magnetic field *(electromagnetic deflection).* Most oscilloscopes use electrostatic deflection, as it permits higher-frequency operation and requires negligible power. Electromagnetic deflection is common in TV picture tubes.

The principles of electrostatic deflection are rudimentary. Because electrons are negatively charged particles, they are attracted by a positive charge or field and repelled by a negative one. Since the electron beam is a stream of electrons, a positive field will divert it in one direction, and a negative field will divert it in an opposite direction. To move the beam in this way in the CRT, *deflecting plates* are mounted inside the tube, as shown in Fig. 1-3, and suitable deflecting voltages are applied to them.

These plates are arranged in two pairs: H_1 and H_2 for deflecting the beam horizontally, and V_1 and V_2 for deflecting it vertically. Leads are

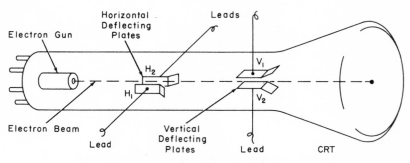

Fig. 1-3. The arrangement of the deflecting plates.

attached for external connections. The beam passes down the tube between the four plates. When the plates are at zero voltage, the beam is midway between them, and the spot is in the center of the screen (Fig. 1-4A). When H_1 is made positive with respect to the cathode (all other plates at zero voltage), it attracts the beam, and the spot moves horizontally to the left (Fig. 1-4B); when H_2 is made positive, the beam is attracted toward that plate, and the spot moves horizontally to the right (Fig. 1-4C). Similarly, when V_1 is made positive, it attracts the beam, and the spot moves vertically upward (Fig. 1-4D); when V_2 is made positive, the beam is attracted toward that plate, and the spot moves vertically downward (Fig. 1-4E). In each of these deflections, the displacement of the beam (and therefore the distance traveled by the spot) is proportional to the voltage applied to the plate.

If a negative voltage is applied to any plate, the beam will be repelled, rather than attracted, and the deflection will be in a direction opposite to that indicated in the preceding paragraph. Thus, when V_2 is made negative, it will repel the beam, and the spot will move vertically upward.

As was mentioned in Section 1.2, when a spot moves too rapidly for the eye to follow, it traces a line. This is what happens in the CRT when a rapidly pulsating or alternating voltage is applied to the deflecting plates; the beam is flipped back and forth so rapidly that the spot traces a line. When a positive pulsating voltage is applied to H_1 (or a negative pulsating voltage to H_2), the spot traces a horizontal line from center to left (Fig. 1-4F); when the positive pulsating voltage is applied to H_2 (or negative pulsating voltage to H_1), the spot traces a horizontal line from center to right (Fig. 1-4G). Similarly, when the positive pulsating voltage is applied to V_1 (or negative pulsating voltage to V_2), the spot traces a vertical line from center upward (Fig. 1-4H); when the positive pulsating voltage is applied to V_2 (or negative pulsating voltage to V_1), the spot traces a vertical line downward (Fig. 1-4I).

When an alternating deflection voltage is applied to H_1 or H_2, the spot moves alternately from center to one side and back, and from center to the opposite side and back—tracing a line that passes through the center of the screen. (This results from the alternate attraction and repulsion of the

First Principles of Oscilloscopes 5

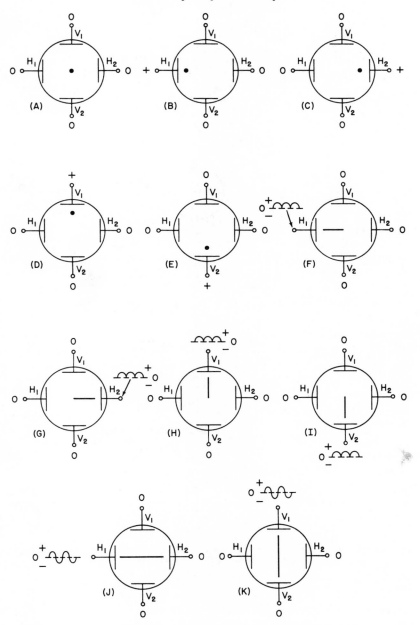

Fig. 1-4. Deflection when the voltage is applied to one axis.

beam by the positive and negative ac half-cycles.) Thus, a horizontal line is traced when an ac voltage is applied to either horizontal plate (Fig. 1-4J); a

6 *Practical Oscilloscope Handbook*

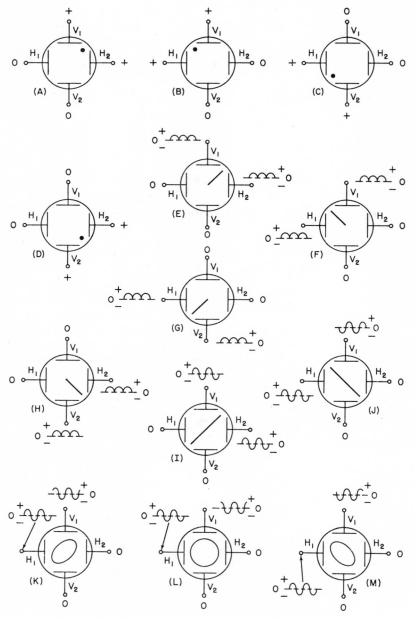

Fig. 1-5. Deflection when the voltage is applied to both axes.

vertical line is traced when an ac voltage is applied to either vertical plate (Fig. 1-4K).

All of these examples show what happens when a voltage is applied to the horizontal *or* vertical deflecting plates. When a horizontal and a vertical voltage are applied simultaneously, deflection of the beam is proportional to the resultant of the two voltages, and the position of the spot is intermediate between the horizontal and vertical axes of the screen. This is illustrated by the patterns in Fig. 1-5.

In Fig. 1-5A-D, a steady positive voltage is applied to one horizontal plate and one vertical plate. When these two deflecting voltages are equal, the position of the spot is 45° from the horizontal axis, as shown. The angle is proportionately greater than 45° (spot closer to vertical axis) when the vertical voltage is the higher of the two, and is less than 45° (spot closer to horizontal axis) when the horizontal voltage is the higher. When two negative voltages are used, deflection is in the opposite direction.

If, instead of a steady voltage, a pulsating positive voltage is applied to the same plates as in the preceding example, the patterns in Fig. 1-5E-H will be obtained. Here, as before, the tilt of the trace is 45° from horizontal when the two voltages are equal and in phase. The tilt is greater than 45° (outer tip closer to vertical axis) when the vertical voltage is the higher of the two, and is less than 45° (outer tip closer to horizontal axis) when the horizontal voltage is the higher. When a negative pulsating voltage is applied to both plates, the trace extends in the opposite direction.

When an alternating voltage is applied to the plates *in phase*, the patterns of Fig. 1-5I-J are obtained. As in the preceding example, the tilt of the trace is 45° from the horizontal when the two voltages are equal. The tilt is greater than 45° (tips closer to vertical axis) when the vertical voltage is the higher of the two, and is less than 45° (tips closer to horizontal axis) when the horizontal voltage is the higher. Actually, two more patterns (the ac equivalent of Fig. 1-5G and H) are obtained when ac voltages are applied to H_1 and V_2, and H_2 and V_2, respectively, but to the eye these patterns are the same as those of Fig. 1-5I and J. The ac trace has the same length from center-screen to each tip when the ac is symmetrical; when the ac is asymmetrical, the shorter half of the trace corresponds to the lower-voltage half-cycle.

A single-line trace is obtained only when the phase angle between the horizontal and vertical ac voltages is 0°, 180°, or 360°. At other phase angles, a double-line trace is obtained: at equal voltages the pattern becomes an ellipse with right tilt (Fig. 1-5K) for angles between 0° and 90°; a circle at 90° (Fig. 1-5L); an ellipse with left tilt (Fig. 1-5M) between 90° and 180°; again an ellipse with left tilt between 180° and 270°; again a circle at 270°; and again an ellipse with left tilt between 270° and 360°. At 0° and 360°, the single line tilts right (Fig. 1-5I); at 180°, it tilts left (Fig. 1-5J).

1.4 CRT Features

Electrostatic cathode-ray tubes are available in a number of types and sizes to suit individual instrument requirements. A brief discussion of the important features of these tubes follows. Detailed electrical and mechanical data may be found in tube manufacturers' tables.

Size. Size refers to screen diameter. CRTs are available in 1-, 2-, 3-, 5-, and 7-inch sizes for oscilloscope use. The 5-inch size is the most common for stationary and general-purpose oscilloscopes; the 3-inch size is the most common for portable instruments. The first figure in the type number usually expresses the size. Thus, the 5GP1 is a 5-inch tube. Both round and rectangular CRTs are found in scopes today. The vertical viewing size of a rectangular CRT is 8 cm and the horizontal axis is 10 cm.

Phosphor. The screen is coated with a fluorescent chemical called a *phosphor.* This material determines the color and persistence of the trace, and the phosphor number indicates both. The trace colors in electrostatic CRTs for oscilloscope use are blue, blue-green, and green (white is used for television; blue-white, orange, and yellow are used for radar). Persistence in these tubes is expressed as *short, medium,* and *long*; this refers to the length of time the trace remains on the screen after the signal has ended.

The phosphors of oscilloscope tubes are designated as follows: P1, green medium; P2, blue-green medium; P5, blue very short; and P11, blue short. These designations are combined in the tube type number. Thus, the 5GP1 is a 5-inch tube with a medium-persistence green trace.

Medium persistence trace suits most general-purpose applications. Long persistence is needed for transient studies, since it keeps the image on the screen for observation for a short period after the fast transient has disappeared. This persistence is invaluable, also, for displaying very slow phenomena which otherwise would only produce a slow-moving dot on the screen. Short persistence is needed for extremely high-speed phenomena, to prevent the smearing and interference caused when one image persists into the period of the next one. Phosphor P11 is considered to be the best for photographing from the CRT screen.

Operating Voltages. The CRT requires a heater voltage (commonly 6.3 V ac or dc at 600 mA, but 2.5 V at 2.1 amp in a few types) and several dc voltages which depend upon tube type. The latter include:

1. Negative grid (control electrode) voltage, -14 to -200 V.
2. Positive anode no. 1 (focusing electrode) voltage, 100-1100 V.
3. Positive anode no. 2 (accelerating electrode) voltage, 600-6000 V.
4. Positive anode no. 3 (accelerating electrode in some types) voltage, 200-20,000 V.

Deflection Voltages. Either an ac or dc voltage will deflect the beam (see Section 1.3). The distance through which the spot moves on the screen is proportional to the dc or peak ac amplitude.

The deflection sensitivity of the tube usually is stated as the dc voltage (or peak ac voltage) required for each inch of deflection of the spot on the screen. (Sometimes this is given in rms volts, and sometimes in centimeters deflection.) Sensitivity is stated separately for the horizontal and vertical plates. It varies between 26 and 310 dcV/in. for horizontal, and between 18 and 350 dcV/in. for vertical, depending upon tube type.

Complex Tubes. Some oscilloscope CRTs have more than one electron gun. This permits the simultaneous display of signals on the single screen. Such tubes are available with two to four guns but are not common. Operating and deflecting voltages are applied separately to each gun; each one has separate controls for focus, brightness, and other adjustments. Dual-beam and dual-trace scopes are described in detail in Section 1.13.

1.5 Viewing Screens

It is hard to denote the exact position of the spot on the plain face of the CRT. A properly graduated transparent screen, therefore, is placed in front of the tube for accurate measurement and observation of spot position.

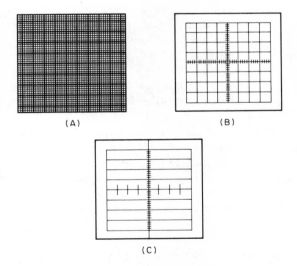

Fig. 1-6. Viewing screens.

Such screens (known variously by the terms *graticule, grating, grid, grid screen, mask,* and *screen*) take several familiar forms. The principal ones are shown in Fig. 1-6. The major divisions on these screens usually are either centimeters or half-inches. Some screens, especially those intended for radio and television servicing, have markers to indicate a calibrating-voltage deflection, such as 1-volt peak-to-peak.

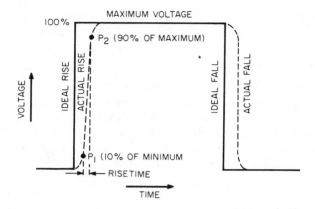

Fig. 1-7. Risetime is measured from the 10 to 90 percent point.

1.6 Function of Amplifiers

Some of the voltages listed under *Deflection Voltages* in Section 1.4 are high. A CRT vertical sensitivity of 50 dcV/in., for example, means that 150 volts dc or peak ac will be needed to move the spot three inches on the screen. Such a strong signal is seldom available, so weak ones must be raised to this level. Deflection amplifiers are used for this purpose in an oscilloscope.

An oscilloscope has separate horizontal and vertical amplifiers. The frequency response of these amplifiers must be wide enough to accommodate faithfully the entire band of frequencies handled by the oscilloscope. The passband of a high-quality laboratory instrument extends from dc to a frequency of 30 to 1000 MHz; a less costly television service oscilloscope must have a 5-MHz passband. An oscilloscope intended for audio testing usually handles ac only and has an upper limit of 200 kHz. The sensitivity (gain) of oscilloscope amplifiers affords deflection of 20 to 200 mV/in. in low-cost instruments, and 0.5 mV/in. or better in high-priced ones.

Most oscilloscopes provide direct access to the deflection plates, so that a high-voltage signal, when available, may be applied to the plates without unnecessarily going through an amplifier. A critical evaluation of amplifier performance involves rise time, the time required for the output to vary from 10 to 90 percent of its final value when a step function (square wave) is applied to its input (see Fig. 1-7). It is a proper measure for an oscilloscope since its specific function is to display events, such as voltage variations, in an instantaneous manner without distortion.

In general, if a square wave is applied to the vertical input of a scope, with its associated amplifiers and attenuators, one of the following outputs could be expected:

1. Ringing or oscillations at various frequencies may appear, causing the waveform at Fig. 1-8A to be displayed.
2. Circuit or stray capacitances may affect the high-frequency components of the input square wave, resulting in a rounded-off, rather than sharp, leading edge (Fig. 1-8B).
3. A well-designed input system would produce a response with a slight ramp or tilt of the leading edge of the square wave (Fig. 1-8C). There would be no ringing or oscillation to indicate the addition of signal content not present in the original input signal or any rounding off or indication that some signal components were lost.

With many applications of scopes now directed at digital pulse waveforms, risetime is a critical performance characteristic. And risetime is closely related to bandwidth. Thus, if it is necessary to view particularly narrow, short-duration pulses, it is possible to relate the risetime demand with the bandwidth specification of the scope.

To convert risetime to bandwidth, simply divide 0.35 by the risetime, in seconds, to find the bandwidth in Hz. For example, a 10-nanosecond (nsec) or 10^{-9}-second risetime pulse would require a 35-MHz bandwidth to display its waveform properly; a sharper pulse, with a risetime of 2 nsec, would require a scope whose vertical amplifier bandwidth was at least 175 MHz.

1.7 Function of Sweep Generator

The amplitude of a voltage may be directly measured on a calibrated viewing screen from the length of the straight-line trace it produces (Fig. 1-

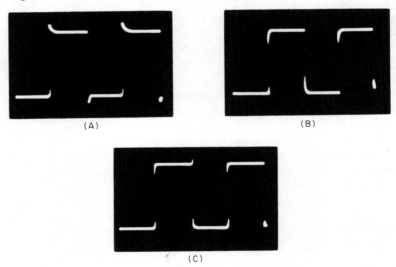

Fig. 1-8. Square-wave response curves.

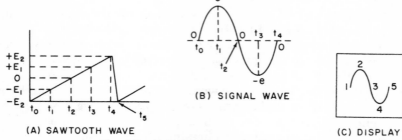

Fig. 1-9. Details of the sawtooth sweep.

4F-K). This is entirely satisfactory for a dc voltage, but the straight line tells little or nothing about the waveform of an ac voltage, pulsating voltage, or transient. What is needed is a graph of the voltage, traced on the screen by the spot.

To obtain such a display, the signal voltage is applied to the vertical plates (directly or through the vertical amplifier) and it moves the spot vertically to positions corresponding to the instantaneous values of the signal. At the same time, the spot is deflected horizontally across the screen by a *sweep voltage* applied to the horizontal plates. The combined action of these two voltages causes the spot to trace out the signal: the horizontal sweep voltage provides the time base by moving the spot horizontally with time, while the signal voltage moves the spot vertically in proportion to the voltage at a particular instant in time.

There are two important sweep-generator requirements: (1) The sweep must be linear—the sweep voltage must rise linearly to the maximum value required for full-screen horizontal deflection of the spot. (2) The spot must always be swept in one direction (normally, left to right) but not in the "return" direction (right to left)—if this is not done, the signal will be traced backwards during the return sweep. This means that the sweep voltage must drop suddenly after reaching its maximum value. These requirements call for a sweep voltage having a linear sawtooth waveform (see Fig. 1-9A).

The manner in which this sawtooth sweep operates is indicated in Fig. 1-9A-C. At time t_0, the sawtooth voltage is $-E_2$; this negative horizontal voltage moves the spot to point 1 on the screen (Fig. 1-9C). At this instant, the signal voltage is zero (t_0 in Fig. 1-9B), so the spot rests at the left end of the zero line of the screen. At time t_1, the linearly increasing sawtooth reaches $-E_1$, which, being more positive, moves the spot to screen point 2. At this instant, the signal voltage is e, the positive peak value, so point 2 is the maximum upward deflection of the spot. At time t_2, the sawtooth voltage is zero, there is no horizontal deflection, and the spot is at center screen (point 3). At this instant, the signal voltage is zero, so there is no vertical deflection either. At time t_3, the sawtooth voltage is $+E_1$, moving the spot to point 4. At this instant, the signal is $-e$, the negative peak value, so point 4 is the maximum downward deflection of the spot. At time t_4, the sawtooth voltage is $+E_2$, moving the spot to point 5. At this instant, the signal voltage is zero, so the spot is not vertically deflected. Between t_4 and t_5, the sawtooth voltage quickly falls through zero to its initial value of $-E_2$, snapping the spot back to position 1 in time to sweep forward on the next cycle of signal voltage.

When sweep and signal frequency are equal, a single cycle appears on the screen; when sweep is lower than signal, several cycles appear (in the ratio of the two frequencies); when sweep is higher than signal, less than one cycle appears. The display is stationary only when the two frequencies are either equal or in integral multiple relationship. At other frequencies, the display will drift horizontally.

Sawtooth sweep voltage is generated by a multivibrator, relaxation oscillator, or pulse generator. The upper frequency generated by internal devices in the oscilloscope is 50 to 100 kHz in audio instruments, 500 to 1000 kHz in television service instruments, and up to several megacycles in high-quality laboratory instruments. In some oscilloscopes, the sweep is calibrated in Hz and kHz; in others it is calibrated in time units (microseconds, milliseconds, and seconds).

Recurrent Sweep. When the sawtooth, being an ac voltage, rapidly alternates, the display is repetitively presented, so that the eye sees a lasting pattern. This repeated operation is termed *recurrent sweep*.

Single Sweep. The opposite of recurrent sweep is single sweep. The latter produces one sweep of the spot across the screen in response to a trigger signal which may be initiated by the signal under study or by means of a switch.

Driven Sweep. The sweep generator is said to be free-running when it operates independently. A hazard of this type of operation, in some applications, is the chance that the sweep cycle will start *after* the signal cycle, thereby missing a part of the signal. Driven sweep removes this possibility because it is initiated by the signal; the signal cycle and sweep cycle therefore start in step. Both driven single sweep and driven recurrent sweep are available.

Non-sawtooth Sweep. In some applications, especially where waveform is of no interest, linear sweep is not needed. In such instances, sine waves or other shapes may be used for sweeping. The patterns in Fig. 1-5K-M, for example, are produced when a sine-wave sweep is used with a sine-wave signal of the same frequency.

Triggered Sweep. There are two basic types of sweeps: recurrent (or free-running) or triggered. In the recurrent mode the voltage rises to a maximum, drops to a low level, then repeats this pattern over and over again. The electron beam moves slowly from left to right, retraces rapidly to the left, and repeats this pattern. Whether an input signal is applied to the scope or not, the horizontal sweep action takes place, and a horizontal line will be displayed on the scope screen.

A triggered sweep, by contrast, does not start until initiated by a trigger voltage, generally derived from an incoming signal. In the absence of an input signal, the sweep is held in the "off" state and the cathode-ray tube is blanked "off" so nothing appears on the screen.

The recurrent sweep approach uses a free-running multivibrator which covers a wide frequency range and can be "locked" into synchronization by an input signal. Synchronization takes place at the signal frequency or a submultiple. For many applications, adequate performance results. A triggered scope does not use a continuous or recurrent sweep. Instead it uses a monostable multivibrator which is in its "off" state until a trigger pulse arrives; thus there is no horizontal deflection on the screen. When an input signal is applied, a trigger pulse is generated and applied to the multivibrator, which switches "on" and produces a sweep signal and a trace. At a specific voltage level, corresponding to the CRT beam arriving at the right-hand side of the screen, the multivibrator switches back to its "off" state, causing the beam to move rapidly to the left-hand side of the screen. A blanking circuit will keep the beam off until sweep is again initiated. The sweep signal produced by this design is very linear and can be calibrated in "time per cm (or inch)" to permit accurate time and frequency measurements directly from the scope.

A significant difference between recurrent and triggered scopes is that the recurrent sweep locks at a frequency of the input signal; the triggered scope displays a trace for a specific period of time. Thus, the triggered scope is on during a specific time interval and will display a waveform, or segment of a waveform, regardless of the signal frequency. One-shot pulses, such as transients or single clamped oscillations, can be observed on a triggered scope, not on a recurrent type.

Many triggered scopes can be operated in three modes: manual triggered, auto triggered, or free-running. In the auto (automatic) mode, the trigger or sync level is preset since the scope's external sync level control does not require any adjustment. As soon as an input signal is applied, all the operator has to do is set the input sensitivity and desired sweep speed and a stable pattern will be displayed. It is common to include circuitry so that a trace is shown on the CRT whether a signal is present or not, in the auto mode.

In the manual mode, it is possible to set the sync level control manually so the scope triggers on any portion of the input signal desired. This is very useful when it is necessary to inspect a small section of the input waveform.

The free-running mode converts the scope's sweep to free-running, allowing it to lock in on a particular frequency or sub-multiple; this is handy when servicing the sync or sweep section of TV receivers.

A very convenient feature on triggered scopes is the calibrated sweep speed, in "time per cm or division." Sweep frequency is the reciprocal of

First Principles of Oscilloscopes 15

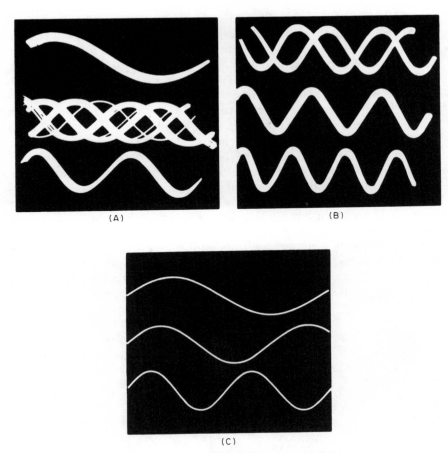

Fig. 1-10. Locking a recurrent sweep scope.

the time period; thus if the sweep speed or time base is set for one millisecond (or msec) per division, and the horizontal line on the scope includes ten divisions or 100 milliseconds, the sweep frequency would be 1 ÷ 100 msec or 100 Hz. A time base of 0.5 microseconds per division would represent 10 × 0.5 or 5 microseconds for a frequency of 200 kHz.

For example, if it was desirable to inspect the horizontal sync pulse in a TV receiver, it must be known that the horizontal sync pulses occur every 63.5 μsec (microseconds). Thus, the sweep speed on a triggered scope would be set at 10 μsec per division, assuming 10 divisions or 100 μsec to cover the screen's horizontal trace. The TV receiver's video signal, horizontal sync pulse, and color sync would then be displayed.

The time base on a scope is a specification that is critical in the selection of a scope. Here's why. It is generally desirable to observe 2 to 4 cycles of a waveform under examination. Thus, if a power supply is being tested for hum, it is common to set the sweep speed to the equivalent of 30 Hz; any 60-Hz hum signal will be displayed as two consecutive waveforms on the screen. If the ripple or hum frequency is 120 Hz for a full-wave rectifier, the sweep rate would be changed to 60 Hz. Fine. Now, suppose it is desirable to observe the 3.58-MHz color sync signal from a color TV set. If a scope has sufficient bandwidth in the vertical amplifier to handle a 3.58-MHz signal, there still may be a problem if the scope does not offer a sufficient sweep speed. Suppose the scope's highest sweep rate is 100 kHz; rather than the display of 2 to 4 cycles of the color sync signal, 35 narrow color sync pulses would be crammed onto the scope screen. If the scope had a sweep speed setting of 1 MHz or 0.1 μsec per division, four cycles could be viewed over the width of the screen.

In a recurrent sweep scope, the free-running oscillator can be locked to the frequency or sub-multiple of the signal under observation. As shown in Fig. 1-10A, one complete sine wave of a signal is displayed. As the sweep frequency control is rotated, a blur of multiple waves appears but they are unstable (Fig. 1-10B); as the control is further rotated, two complete sine waves will become locked in the CRT display as in Fig. 1-10C. The two sine waves are in sync when the horizontal sweep rate is exactly one half of

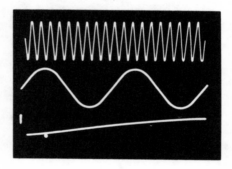

Fig. 1-11. Locking a triggered sweep scope.

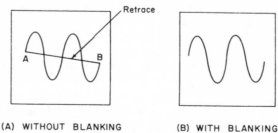

(A) WITHOUT BLANKING   (B) WITH BLANKING

Fig. 1-12. The effect of retrace blanking.

the signal frequency. If the sweep rate is further lowered, more sine waves can be locked in at each sub-multiple of the input frequency. However, with recurrent sweep, not less than one complete cycle can be displayed. Therefore, if it is desirable to view only a small portion of a series of pulses in one particular cycle, use of a recurrent sweep scope is not the solution.

A triggered scope, on the other hand, sweeps for a particular period of time after it is triggered on; therefore any portion of an input signal can be displayed even though it may be one half, one fifth, or three times the input frequency. Thus, triggered scopes can be used to observe small segments of a complex waveform with ease (see Fig. 1-11).

1.8 Synchronization

Sweep frequency drift causes an unsteady pattern; this is seen in pattern migration across the screen and in a change in the number of cycles displayed.

Drift is eliminated by synchronizing the sweep generator with another frequency source (external sync) or with the signal itself (internal sync). The generator may be synchronized from either the positive or negative half-cycle of sync voltage, which are usually selected with a switch on the front panel of the oscilloscope. Sync voltage is injected at a suitable point in the sweep generator circuit.

1.9 Blanking

During the brief *flyback* interval (t_4 to t_5 in Fig. 1-9A), when the sweep voltage snaps from its final amplitude back to its initial value, the spot backtracks across the screen to the starting point. This action can trace an extraneous line (the *retrace*) across the display, as shown from B to A in Fig. 1-12A.

The retrace inserts a false line in the pattern, which causes confusion and inaccuracy. In modern oscilloscopes, the retrace is eliminated by darkening the screen during the flyback; the beam is cut off during this interval. This is done by momentarily applying a negative cutoff voltage (generated by the flyback) to the control electrode. Figure 1-12A shows the pattern defaced by the retrace; Fig. 1-12B shows the pattern with blanking.

1.10 Intensity Modulation

In some applications, an ac signal is applied to the control electrode of the CRT. This causes the intensity of the beam to vary in step with the signal alternations. As a result, the trace is brightened during positive half-cycles and darkened or blanked out during negative half-cycles.

This process, called *intensity modulation* or *Z-axis modulation* (in contrast to horizontal *X-axis* and vertical *Y-axis*), produces bright segments or dots on the trace in response to positive peaks, or dim segments or holes in response to negative peaks.

Intensity modulation is invaluable in many test procedures in which signals are applied simultaneously to the control electrode and one or both deflecting plates.

1.11 Oscilloscope Basic Layout

The simplest oscilloscope consists of CRT, power supply, focus control, and intensity control. Such an arrangement is sometimes used in applications, such as transmitter modulation checking from trapezoidal patterns, where high signal voltages are available for the deflecting plates and special sweep is not needed. Most oscilloscopes, however, are more complex than this, since their applications call for such complements as amplifiers, sweep generators, synchronizing circuits, and a calibration-voltage source.

To describe all combinations which are possible, and which in fact are found, in commercial oscilloscopes is beyond the scope of this book. Figure 1-13 shows the layout of the conventional *complete* instrument. This block diagram gives essential sections and channels; special-purpose oscilloscopes contain additional sections.

The CRT heater and the heaters of tubes in the amplifiers and sweep generator are powered by the ac filament supply (A). This is usually nothing more than the filament windings on the main power transformer of the instrument. This section also supplies an accurate ac calibrating voltage to terminal 1. CRT dc voltage is obtained from the high-voltage dc supply (B) through a voltage divider string, R_1 to R_5, inclusive. Note that R_3 in this string is a potentiometer for varying the focusing-electrode voltage and is the *focus control,* and that R_5 is a potentiometer for varying the control-electrode voltage and is the *intensity control.* Capacitor C_1 grounds the deflecting plates and the second anode for the signal voltage but dc isolates these electrodes from ground. A somewhat different arrangement is used in a dc oscilloscope. (Some instruments do not operate the deflecting plates against ground in this manner, but use push-pull amplifiers to excite both plates of each pair.) The low-voltage supply (G) provides dc voltages for the amplifiers and sweep generators. Often, this latter supply is integral with high-voltage supply B.

Block C represents the vertical amplifier accessible through front-panel input terminal 5; D, the horizontal amplifier accessible through front-panel input terminal 6; E, the linear sawtooth sweep generator; and F, the sync amplifier accessible through front-panel input terminal 7. Although a single ground (common) input terminal is shown as 8 for simplicity, separate ground terminals are usually provided for the horizontal and vertical inputs. In addition to the amplifier input terminals, separate input terminals (3 and 4) provide direct input to the vertical and horizontal deflecting plates, respectively. Also, terminal 2 provides direct access to the control electrode for intensity (Z-axis

First Principles of Oscilloscopes 19

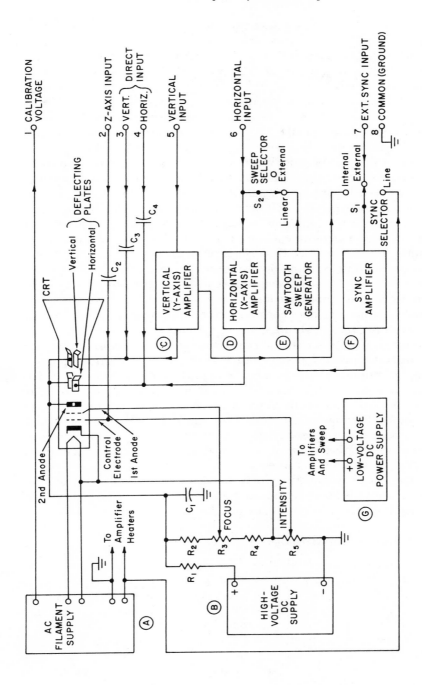

Fig. 1-13. The basic layout of an oscilloscope.

modulation. (In some oscilloscopes, there is also a Z-axis amplifier.) Capacitors C_2, C_3, and C_4 provide dc blocking.

Normally, switch S_2 is set to its LINEAR position. This connects the sweep generator output to the horizontal amplifier input. The sweep voltage accordingly is amplified before being applied to the horizontal deflecting plates. When an externally generated sweep is desired, S_2 is thrown to its EXTERNAL position, and the external sweep generator is connected to input terminal 6. The sweep synchronizing voltage is applied to the internal sweep generator (E) through switch S_1, which permits selection of the type of sync. When S_1 is set to its EXTERNAL position, sync is obtained, through sync amplifier F, from an external sync signal source connected to input terminal 7. When S_1 is at INTERNAL, the test signal itself, entering the oscilloscope at vertical input terminal 5, is used to synchronize the sweep. When S_1 is at LINE, a low voltage of power-line frequency is taken from the ac supply (usually from one of the filament windings of the transformer) to synchronize the sweep.

This simplified layout does not show gain controls of the various amplifiers, sweep frequency controls, sync polarity selector, beam centering controls, or astigmatism control. Presentation of the basic layout of the complete oscilloscope does not require their inclusion. The function and adjustment of all controls are given in the next chapter.

Signals may be applied to the vertical (Y) axis through the vertical amplifier via input terminal 5, or directly to the vertical deflecting plates via input terminal 3. Signals may be applied to the horizontal (X) axis through the horizontal amplifier via input terminal 6, with S_2 set to EXTERNAL to disable the internal sweep generator, or directly to the horizontal deflecting plates via input terminal 4. Signals may be applied to the Z-axis for intensity modulation via input terminal 2. A synchronizing signal may be applied to the internal sweep generator via input terminal 7 and the internal sync amplifier.

The calibrating voltage may be applied to the vertical axis by running a lead between terminals 1 and 5, or to the horizontal axis by running a lead between terminals 1 and 6.

1.12 Plug-in Oscilloscope

A conventional oscilloscope includes all of its basic sections, excluding probes. A plug-in oscilloscope consists of a mainframe, to which a variety of specialized subsystems can be applied (Fig. 1-14). In addition to such plug-in units as vertical and horizontal amplifiers and calibrated time bases, other laboratory functions such as a frequency counter, spectrum analyzer, digital voltmeter, and logic analyzer are also available.

Let's first consider the vertical amplifier plug-ins. A basic plug-in unit would include ac/dc coupling, calibrated attenuators, and vertical

Fig. 1-14. Mainframe with plug-in amplifiers and sweep sections. *(Courtesy,* Hewlett-Packard.)

positioning, and would be capable of handling signals over a wide dynamic range. A high-sensitivity plug-in could replace the basic plug-in if sensitivities as low as 10 μV were under examination. A dual-trace plug-in could be inserted in the mainframe should it become necessary to view and compare separate signals; depending on the need, dual-trace plug-ins are available with individual or separate triggering, with sequential triggering, and with the ability for each channel to have gain and positioning separately set.

A differential plug-in is available to measure and display the difference between two signals, with whatever signal is common to both (common-mode) rejected. If an application involves the display of signals that are hidden by other interfering signals, plug-ins are available with adjustable filters that can be set to attenuate the interference and thus display only the desired signals.

Most horizontal plug-ins are time base units, which provide sweep time ranges from microseconds to seconds, with options for magnification. Also included are multiple provisions for triggering to handle a large range of input signal variations. Also available within horizontal plug-ins are delay lines for delayed-sweep applications.

The advantage of a plug-in oscilloscope is fairly obvious. It is advisable to purchase such a scope when a laboratory or test setup is to handle a large number of applications requiring the scope. As new and more complicated applications arise, it is not necessary to set aside or scrap this initial investment as might be the case if a particular model of a conventional scope had been purchased. Instead, from the wide variety of plug-ins available, the mainframe could be upgraded and the initial investment in the oscilloscope could be maintained. For a laboratory or production setup where a variety of new applications and thus new demands is likely, a plug-in scope offers a fine, long-range selection.

However, in a service shop, for example, it is more likely that the applications for the scope will not vary considerably over a long period of time. In such a case, the choice of a plug-in model, with its many available options, might be extravagant since a conventional 5- or 10-MHz scope would probably supply the full range of needs required in a service shop.

1.13 Dual-Trace Oscilloscope

There are a considerable number of conventional, single-trace scopes used in industry and service shops; they are well suited for a large number of applications. But there are increasing demands for dual-trace scopes to compare time, phase, pulse width, and amplitude of two waveforms in a given circuit or subsystem (Fig. 1-15). Computer hardware, jammed with precisely timed digital signals, are ideally serviced with a dual-beam or dual-trace scope.

A dual-beam scope is basically two scopes in one envelope or tube structure. Inside the envelope are two electron guns and two sets of deflection plates. There are two basic disadvantages of a dual-beam scope. Since the two electron guns cannot be placed at the same point (the concentric area within the neck of the CRT), there will be deflection distortion from each gun since neither beam will strike the center of the tube when deflection voltage is zero. Second, a dual-beam CRT is much more expensive to produce than a single-gun CRT. However, there are several advantages of a dual-beam scope over a dual-trace: (1) both viewed waveforms are exactly in proper phase since they are scanned by a single sweep and triggering system; (2) both traces are brighter since there is no time-sharing of a single electron beam; and (3) there are no disadvantages of chop interference or flicker sometimes encountered with dual-trace scopes.

A dual-trace scope avoids the disadvantages of a dual-beam tube by using a single-gun CRT in conjunction with a time-sharing arrangement. Two display modes are used for dual-trace scopes: dual-alternate and dual-chopped. In dual-alternate mode, the first signal to be observed is routed to channel A and the second signal to channel B; two separate vertical amplifiers are used, each with its own attenuators, position control, and gain controls. On the first cycle of sweep, channel A is displayed. During

First Principles of Oscilloscopes 23

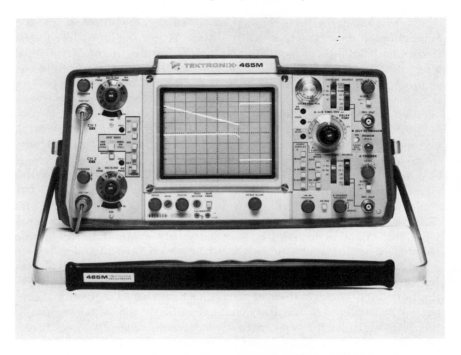

Fig. 1-15. Two signals displayed simultaneously with a dual-beam scope. *(Courtesy,* Tektronix.*)*

the next sweep, channel B is displayed (Fig. 1-16). The third sweep cycle would display channel A, and the process would continue. Since channel A and channel B have their own separate vertical position controls, it is simple to adjust the display so that channel A is placed above or below

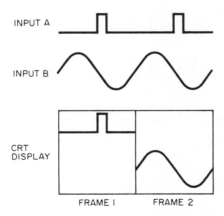

Fig. 1-16. Dual-alternate mode.

channel B; similarly, each channel has its own gain control so that the height of each signal can be individually set. Triggering for both channels is selected from either channel A or channel B; thus, channel A may be selected as the trigger source, and channel B will then be triggered by the channel A signal. With a triggered scope (most dual-trace scopes are triggered), it is then simple to measure the phase or real-time difference between input signals A and B.

Alternate mode is convenient except at low sweep rates, when it becomes possible to see first one trace appear and then the other. Under these conditions, a dual-chopped mode is preferable. Here the single electron beam is not switched at the end of each sweep cycle to the next channel; instead, the electron beam is rapidly switched from channel A to channel B and back and forth at a very rapid rate, generally about 500 kHz (Fig. 1-17). Thus, as the beam is moving from the left to right side of the CRT, the vertical deflection plates are being fed samples of the signal from channel A, then B, then A, etc. Again, each channel includes its own positioning control, attenuator, and gain controls, and thus each display can be individually adjusted for separation and display height. In a sense, the horizontal sweep is divided or chopped into small segments of time shared by channel A and then channel B. Since the switching rate is relatively high and the persistence of the screen is relatively long, there is no distortion or discomfort to the viewer because the dual display would appear as a full pattern for each waveform. The only time a viewing problem arises is when the chop rate is an exact multiple of the chop frequency, and gaps in the waveform may be visible. Should this condition occur, the operator merely selects the dual-alternate mode.

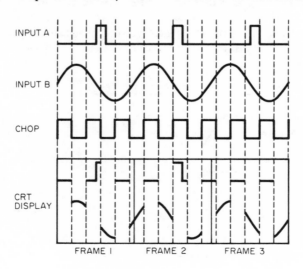

Fig. 1-17. Dual-chopped mode.

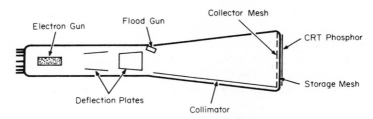

Fig. 1-18. Basic components of a storage scope.

There is one significant limitation to a dual-trace scope: the single trigger and timebase. For example, if it was necessary to display the power-supply ripple (120 Hz) from a TV receiver on channel A and the horizontal sync signal (15,750 Hz) on channel B, the single sweep rate setting would not be adequate. A low-frequency sweep of 60 Hz would be too low for the 15,750 Hz horizontal sync signal, and a 7875-Hz sweep would be too fast for the power-supply ripple waveform. Of course, signals of such widely different frequencies may not be common but it is a conceivable situation.

A number of dual-trace scopes on the market include a feature to switch automatically from the chop to alternate mode when sweep speed is increased beyond a certain point; the scope is switched from alternate to chop when sweep speed is decreased beyond a particular value. Alternate sweep is thus automatically selected for high-speed, short-time sweeps while chop mode is used for low-frequency signals.

There are dual-trace scopes on the market that include switching of triggering as well as vertical amplifier input in the alternate mode. Thus when channel A is being displayed, its sweep is triggered by its input. When channel B is switched on, sync is triggered by the channel B signal.

1.14 Storage Oscilloscopes

When it is necessary to observe short-duration, nonrepetitive waveforms, a conventional scope is barely useful because of its relatively short-persistence phosphor. If a long-persistence phosphor CRT were used in a conventional scope, the display would be too slow to respond to rapid variations of normally viewed waveforms. Thus, there existed a need for a storage CRT, a specially built tube to fill the demand.

Although a number of different types of storage tubes exist, the most popular version is the variable-persistence storage CRT. Basically, as shown in Fig. 1-18, the CRT has the same standard elements of a conventional tube—an electron gun, deflection plates, and a phosphor screen. In addition, the storage CRT includes a flood gun, collimator, collector mesh, and storage mesh. The electron gun, known as the write gun in a storage scope, emits a thin, well-defined beam of electrons at high velocity to the phosphor screen. The flood gun, at a very low potential, emits low-velocity electrons that are directed to the screen. A conductive coating or collimator on the inside of the CRT forces the flood gun electrons to move so that they are parallel to the centerline of the tube. The collector mesh is a fine wire screen mounted parallel to the CRT faceplate and maintained at a voltage slightly more positive than the collimator. The storage mesh, like the collector mesh, is parallel to the CRT faceplate but is coated with a thin layer of insulating material with a high secondary emission characteristic; when this material is bombarded by high-velocity electrons from the write gun, electrons will be emitted leaving the surface area positively charged.

To understand how the storage CRT works, it is necessary to analyze the relative voltages on the various electrodes. The write gun, or standard electron gun, has a cathode voltage of –5kV, the flood gun has a cathode voltage of 0 volts, the collimator potential is about +50 V, the collector mesh +150 V, and the storage mesh is –10 V.

When the high-velocity electron beam leaves the write gun, it is barely affected by the low voltages on the collimator and collector mesh; thus most of the electrons hit the phosphor screen and create a trace. However, some of the high-velocity electrons hit the insulated area of the storage mesh, and high secondary emission in that particular area causes the region to become slightly positive. Now if the input waveform is removed, or does not recur, the write gun is turned off but the trace will remain for a brief period of time. Here's why. Electrons from the flood gun, although at much lower velocity, are attracted by the relatively low positive potential of the collimator and collector mesh to the storage mesh. Because of the –10 V flood gun, electrons will not pass through and reach the phosphor screen. However, in the regions on the insulated material where a positive charge exists, electrons will pass through the storage mesh. Once beyond the storage mesh region, these electrons will be accelerated by the +5kV potential on the phosphor screen coating and cause the phosphor to glow and present a display. The time duration of the storage display is determined by the ability of the storage mesh to hold its charge, generally several minutes.

When the flood gun is turned on to initiate the storage mode, flood-gun electrons collide with gas molecules in the CRT and produce positive ions. These positive ions move to the storage mesh surface, which is charged to –10 V, and gradually reduce this negative charge, thus allowing more and more flood electrons to pass through the storage mesh. After a brief interval, the entire phosphor screen is bombarded by electrons, and thus the displayed waveform is covered or wiped out.

If it is desirable to erase the display quickly, an erase control is available on the scope to place the collector mesh and storage mesh at the same high (+150 V) positive potential for a brief interval, perhaps 200 milliseconds, and then both elements are returned to their original voltage levels. If it is desirable neither to wait the full storage time nor to erase

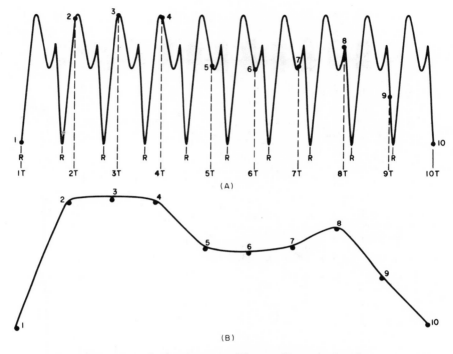

Fig. 1-19. Generating a sampling oscilloscope signal.

rapidly, variable storage time is also available. Instead of applying sudden high potential to the storage mesh, a series of positive-going pulses is applied to the storage mesh. By varying the pulse width and repetition rate, the storate mesh charge will be reduced; a wide pulse width and rapid pulse rate will erase the presentation quickly while a narrow pulse width and low pulse rate will permit the display to remain on the phosphor screen for several minutes.

1.15 Sampling Oscilloscope

When nonrepetitive signals are to be observed, a storage oscilloscope is used instead of a conventional scope. Similarly, when it's necessary to observe repetitive, high-frequency signals beyond the frequency range of conventional scopes, a sampling oscilloscope is employed. A sampling technique converts the high-frequency, repetitive signal to a lower frequency, thus extending the effective bandwidth as high as 18 GHz (18,000 MHz).

Sampling techniques are not new. Suppose a production line manufacturing diodes wished to monitor the quality of the final output. One approach would be to take every single device coming off the production line and test it completely for all parameters—obviously a thorough but

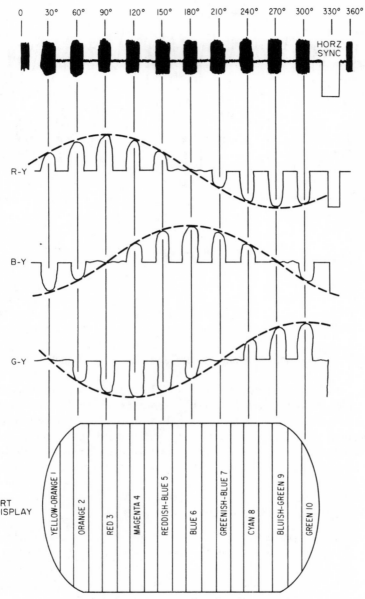

Fig. 1-20. Color-bar signal and CRT display.

rather costly procedure. Another approach for low-cost diode production is to sample every 100th or 1,000th device as it comes off the line. Although testing each diode is certainly more informative than the sampling or spot-checking technique, it is apparent that the sampling approach could do an adequate job of identifying production flaws as they arise and begin to affect the performance characteristics of the diodes.

In a similar manner, many points can be sampled on a high-frequency repetitive signal and then the numerous samples collected to provide a close equivalent of the signal being observed.

Assume it is necessary to observe a 1,000-MHz signal (such as that shown in Fig. 1-19A) from a communications system transmitter. At a time reference point A, on Fig. 1-19A, a sample of the waveform is taken and stored in memory. Then, after the second cycle of the repetitive signal passes the equivalent point A, another sample B is taken and stored in memory. During the third cycle, another sample C is taken, and the process continues until a predetermined number of samples have been taken of the repetitive waveform. Then, all of the samples are taken from the memory to present the waveform shown in Fig. 1-19B. If sufficient samples are taken, the displayed waveform will closely match the original input signal.

1.16 Vectorscope

A vectorscope is a form of dedicated oscilloscope; that is, it is intended strictly for examination of the color TV signal. The display shown on a vectorscope is called a vectorgram and is used to check the phase relationship between the color TV reference signal and the color demodulator's output.

The vectorscope is basically an oscilloscope designed to produce Lissajous patterns. A conventional scope involves an input signal to the vertical axis (where deflection is a function of amplitude) and a time base signal to the horizontal axis (to represent time). A vectorscope, on the other hand, involves two input signals, neither representing time (as ordinarily encountered with Lissajous phase and frequency comparisons).

A color TV set usually includes two color controls on its front panel: hue or tint, for adjusting flesh tones, and color or saturation, for setting the intensity of the color. When a color receiver develops trouble, the technician can service the set using the transmitted color signal as reference and signal trace from one stage to the next. However, if the trouble is elusive, it is more convenient for the technician to use a fixed-pattern signal to troubleshoot rather than the constantly shifting program signal. But, even more significantly, to check and adjust the phase relationship of the color sync circuits, a stable, known reference signal is required. A color bar generator provides a test signal that produces eleven color signals plus one sync pulse, as shown in Fig. 1-20. When the color bar signal is applied to a color receiver, a series of ten vertical bars, ranging from yellow-orange to green, will be observed on the screen, as shown at the bottom of Fig. 1-20. Each color bar and its position on the screen represent a particular phase relationship which is checked by the vectorscope.

Let's review a basic color TV receiver when a color program is received (Fig. 1-21). The brightness or luminance signal is routed to the

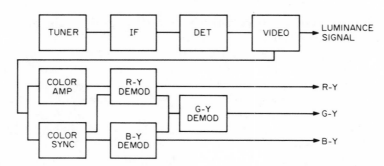

Fig. 1-21. Block diagram of a color TV receiver.

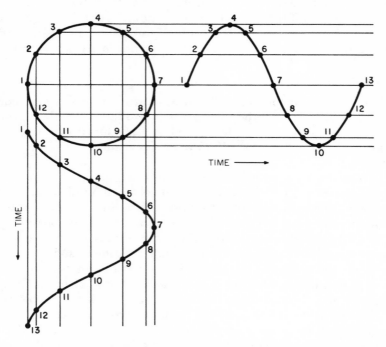

Fig. 1-22. Lissajous pattern of 90° phase difference signals.

video amplifier. The color or chroma information is fed to the color amplifiers and color demodulators (or detectors); the transmitted chroma information is in the form of both an amplitude and phase-modulated signal. Thus, the color demodulators must somehow sense variations in phase as well as signal amplitude. In addition, a section in the color receiver is responsible for accurately reproducing a color reference signal that is a critical factor in modulating the color signals at the transmitter but must be deliberately removed before transmission to prevent interference difficulties. This color sync section accepts a sample "burst" signal sent by the station, compares it with a locally generated signal, and delivers a reference signal at its output that is identical in phase and frequency to the reference oscillator at the station. A check of the operation of this color sync section is performed with the vectorscope.

Now let's review a Lissajous pattern produced when two sine-wave signals, of the same frequency but 90 degrees out of phase, are applied to the horizontal and vertical input of a scope. Assume both vertical and horizontal amplifiers have the same gain. As shown in Fig. 1-22, one complete cycle will develop a circular pattern.

The three color signals, developed at the output of the three demodulators in a color receiver, are R-Y, B-Y, and G-Y (representing the red, blue, and green color voltages). The B-Y and R-Y signals are 90 degrees out of phase with each other; as shown in Fig. 1-20, the third bar represents R-Y while the sixth bar represents B-Y. Other colors are represented by other phase angles relative to R-Y and B-Y. Now, if the color-bar signal, with a series of sharp pulses at the same frequency but displaced by 30-degree differences, is applied to the color receiver, the demodulator outputs will vary. If the B-Y demodulator output is applied to the horizontal input

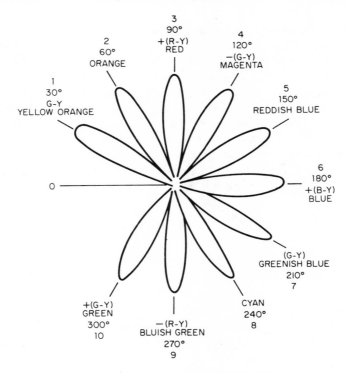

Fig. 1-23. Vectorscope display of color-bar signals.

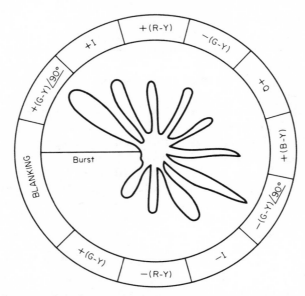

Fig. 1-24. Vectorscope graticule indicates phase angles and colors.

and the R-Y signal to the vertical input of the scope, a pattern similar to Fig. 1-23 will be displayed. This vectorgram is produced by the combination of R-Y and B-Y outputs, and thus a competent service technician can quickly analyze the performance of a color receiver by this display.

As the color control is increased, the amplitude of the R-Y and B-Y outputs should increase, thus enlarging the diameter of the pattern. As the hue or tint control is varied, the third bar or R-Y should shift from its vertical position to a tilted line on either side of vertical.

Vectorscope graticules do not look like a conventional, gridded CRT screen. Instead, they contain a circular pattern with indications for the various colors and phase angle axes used in color receivers, as shown in Fig. 1-24.

2

oscilloscope controls and adjustments

In general, oscilloscope controls may be categorized as *functional controls* (those that are continually used and are usually found on the front panel) and *operating controls* (those used in the calibration and maintenance of the instrument, which usually must not be disturbed by the operator, and generally are found inside the case).

Not all controls will be found in any one type of instrument. For example, a simple, inexpensive oscilloscope will not have an ac–dc switch

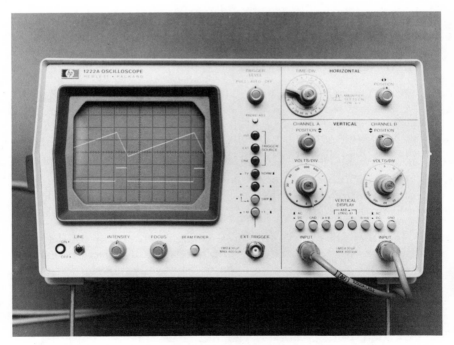

Fig. 2-1. A general-purpose wideband oscilloscope with dual-trace capability. *(Courtesy,* Hewlett-Packard.)

33

if its amplifiers handle ac only. Also, controls and terminals are not always found in the same place (one instrument has the Z-axis input terminal in the rear, whereas another has it on the front panel; direct input to deflecting plates is handled by a selector switch connected to the regular input terminal in one instrument, whereas it is handled by separate rear terminals in another). Expensive, special-purpose oscilloscopes often have more controls because these instruments provide more functions.

Figure 2-1 shows a general-purpose oscilloscope suitable for use by the low-budget laboratory, electronic service technician, student, lecturer, and radio amateur. The bandwidth of its vertical channel is dc to 4.5 megacycles.

Figure 2-2 identifies the front-panel controls and terminals seen in the photograph, and keys these to applicable sections in this chapter.

The important controls and their functions follow:

Screen Illumination. In some oscilloscopes, the engraved lines of the transparent viewing screen are brightened by edge-lighting the graticule. This provides a sharp reproduction of the lines when photographs are made from the screen. A front-panel control is provided for adjusting the brightness of the illumination to suit individual conditions of viewing or photographing, or for extinguishing the light.

Intensity Control. Permits adjustment of trace brightness from total darkness to very bright. Maximum intensity produces a spot or trace that is bright enough to burn the CRT screen permanently if it is allowed to remain in one position too long.

Focus Control. Adjusts trace sharpness.

Horizontal Centering Control. Also called *horizontal position control* or *X-position control.* Adjusting it moves the spot from side to side to any desired horizontal position on the screen.

Vertical Centering Control. Also called *vertical position control* or *Y-position control.* Adjusting it moves the spot up and down to any desired vertical position on the screen.

Amplifier Selector. A switch for changing from ac amplifier to dc amplifier, or vice versa, in the horizontal or vertical deflection channel. Generally, the amplifier is direct-coupled. Basically, therefore, it is a dc amplifier which will also handle ac; the switch inserts (for ac) or short-circuits (for dc) an input coupling capacitor.

Trace Reverser. A polarity-reversing switch which inverts the vertical deflection. In observation of certain signals (such as TV waveforms), the signal is often displayed upside down; it can be righted by throwing this switch.

Oscilloscope Controls and Adjustments 35

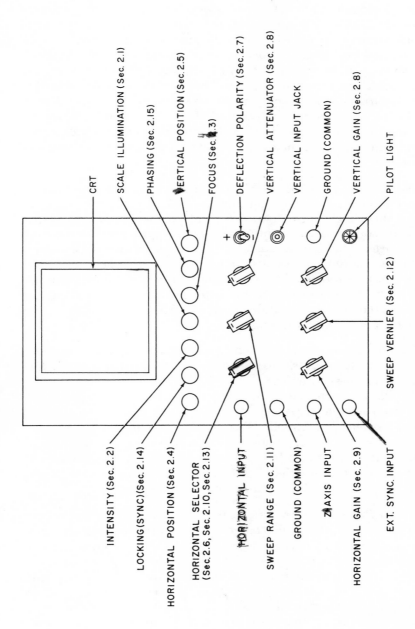

Fig. 2-2. The functional controls of an oscilloscope.

Vertical Gain Control. Called *V-gain (amplitude) control* or *Y-gain (amplitude) control.* Adjusting it, while holding the vertical input signal voltage constant, varies the height of the pattern on the screen.

Some vertical gain controls are more complicated than the single potentiometer. These are *attenuators* which consist of a step-type section and a potentiometer vernier section. The potentiometer provides continuously variable control of gain in any one of the ranges provided by the step-type ("coarse adjustment") section. The steps are integral, usually providing multipliers of ×0.1, ×1, ×10, ×100, and ×1000 for the potentiometer setting. Such attenuators are frequency compensated for wideband response.

Horizontal Gain Control. Also called *H-gain (amplitude) control* or *X-gain (amplitude) control.* Adjusting it, while holding the horizontal input signal voltage constant, varies the width of the pattern on the screen.

Some horizontal gain controls are more complicated than the single potentiometer. These are *attenuators* which consist of a step-type ("coarse adjustment") section and a potentiometer (vernier or "fine adjustment") section. The potentiometer provides continuously variable control of gain in any of the ranges provided by the step-type section. As in the vertical attenuator, the steps are integral, usually providing multipliers of ×1 and ×10 for the potentiometer setting.

Sweep Selector. In simple general-purpose oscilloscopes, the selection includes (1) linear sawtooth (internal), (2) external, and (3) line-frequency sine-wave (internal).

In addition to the selections mentioned in the preceding paragraph, driven sweep, single sweep, delayed sweep, and similar variations are provided by some professional, laboratory-type oscilloscopes.

An oscilloscope designed for television servicing may provide, in addition to the usual sweep-frequency ranges, two preset fixed frequencies for TV alignment. One of these is 30 Hz (TV vertical deflection frequency); the other is 7875 Hz (TV horizontal deflection frequency).

Sweep Range Selector. Often called *coarse frequency control,* permits selection of the frequency range of the internal sawtooth generator.

In professional, laboratory-type oscilloscopes, the sweep ranges often are expressed in time units (microseconds, milliseconds, or seconds per centimeter of screen width) instead of frequency. This is a labor saver, since the time interval of the display, rather than the sweep frequency, is of first importance in many scientific measurements, and time would have to be calculated from sweep frequency if it were not given directly. Typical ranges extend from 0.1 μsec per centimeter to 5 sec/cm. In some instruments, sweep time is stated per scale division, rather than per centimeter.

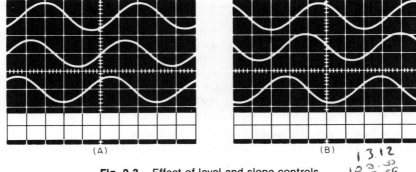

Fig. 2-3. Effect of level and slope controls.

Sweep Frequency Control. Often called *fine frequency control* or *frequency vernier*, permits continuous variation of sweep frequency within any of the ranges provided by the sweep range selector.

Sync Selector. With the sync selector, the operator may select the type of signal used to synchronize the sweep oscillator. The selections generally are (1) + internal; (2) − internal; (3) line-frequency sine wave; and (4) external. (1) uses the positive half-cycle of the vertical input signal; (2) uses the negative half-cycle of this signal; (3) uses a low ac voltage taken from the oscilloscope power supply; (4) uses a signal from an external generator via the self-contained sync amplifier.

Sync Amplitude Control. The gain control of the self-contained sync amplifier, and its adjustment varies the amplitude of the synchronizing voltage applied to the internal sweep generator. At the correct setting of this control, the sawtooth sweep locks in step with the sync voltage, and the display stands still on the screen.

Level and Slope Controls. In a triggered scope, the vertical input signal reaches a particular level, set by the "level" control, and activates a trigger which starts the horizontal sweep. At the completion of sweep, the duration of which is set by the sweep control, the sweep stops and retraces to its starting point; the sweep will not recur until another input exceeds the preset trigger level and develops another trigger pulse. If the level control is set to a level higher than the input signal can reach, triggering will not be possible.

The "slope" switch selects the positive (+)- or negative (−)-going slope of the triggering signal. When the operator selects the positive-going slope, the start or left-hand side of the display will show a signal moving upward or positive. As the level control is rotated, the display will appear to move to the left, as shown in Fig. 2-3A, noting the position of the positive peak of the sine wave. If the slope control is turned to negative-going, the negative portion of the signal will appear at the left of the screen (Fig. 2-3B).

Mode Switch. Triggered scopes, if no input signal or trigger is applied, will not show a display on the screen. This may be confusing to the operator since he may incorrectly assume the scope is not working. Thus an "auto" position is included to allow the sweep to perform as a free-running oscillator and display a horizontal line on the screen with no input signal present.

When an input signal is applied and appears, the mode switch is then set to "manual" to permit locking of the signal and then displaying whatever portion of the signal is desired.

Many scopes, intended for the serious TV technician, include a "TV" position to supply vertical and horizontal sync to the triggering circuits and thus simplify the task of presenting key video and sync waveforms.

Source Switch. Many modern scopes include a source switch to select the triggering input: 60-Hz line, channel 1, channel 2 (dual-trace scopes), or external.

Delay Line. If a sharp pulse is applied to the input of a scope, recurrent or triggered, it takes a finite time for the sweep to begin and the display to appear on the CRT. Thus, the leading edge of a sharp pulse cannot be viewed for analysis.

Since it is not possible to anticipate the arrival of the input signal and trigger before the signal arrives, an alternate solution is to delay the input signal to be displayed. A portion of the input signal is fed to the triggering circuit where it can initiate the sweep; at the same time, the input signal is fed through a delay line, which delays the signal arrival to the vertical amplifier. By the time the delayed vertical signal is amplified and fed to the grid (or cathode) of the CRT, the sweep has already been initiated and thus the leading edge of the pulse can be viewed, as shown in Fig. 2-4.

Sweep Magnifier. It is often desirable to expand the sweep so that a small portion of the waveform can be closely analyzed. A 5× (or 10×) magnifier switch is provided to allow the horizontal amplifier gain to be increased five (or ten) times. The display will then widen by this amount (see Fig. 2-5A and B). Of course, the entire display is now lost since four-fifths (or nine-tenths) of the waveform is off screen. However, by adjusting the horizontal-positioning control, the operator can move the particular segment of the waveform to be displayed to the center area of the CRT.

A disadvantage of this method of sweep magnification is the reduction in display brightness. Since the beam is on the screen only one-fifth (or one-tenth) of its sweep time, there is less time for electrons to bombard the CRT phosphor and develop a bright image.

13. Advance HORIZONTAL GAIN control, noting that spot is deflected into a horizontal line. Note also that length of line is controllable by adjustment of HORIZONTAL GAIN control.
14. Switch off sweep and reduce horizontal gain to zero.
15. Advance VERTICAL GAIN control to mid-range.
16. Touch VERTICAL INPUT terminal, noting that stray signal pickup by hand causes spot to be deflected to give a vertical line. Note also that length of line is controllable by adjusting VERTICAL GAIN control.
17. Reduce vertical gain to zero.
18. Switch on internal sweep and advance HORIZONTAL GAIN control for a horizontal-line trace.
19. Connect any required probe to the VERTICAL INPUT terminals.
20. Oscilloscope now is placed properly into operation and is ready for connection to the test circuit.

If the oscilloscope is a dc instrument with direct-coupled amplifiers, the waiting period in Step 8 will be longer than the prescribed one minute, because the amplifiers drift when started cold and take some time to stabilize. A five-minute period is safe in most cases. Some direct-coupled oscilloscopes continue to drift, at a progressively decreasing rate, up to one or two hours after they have been switched on. Consult the manufacturer's instruction manual for stabilization time.

2.3 How to Set Up a Dual-Trace Oscilloscope

A dual-trace scope offers additional advantages over a single-trace model relative to troubleshooting, waveform comparison, and comparative timing analysis. However, to gain the advantages offered, the operator must become well acquainted with the operating controls—their proper settings and interaction.

Although a dual-trace scope may appear considerably more complicated than a single-trace model, it takes very little time to become fully competent in its use. Almost all recent model dual-trace scopes are fine examples of well-done human engineering design. Vertical amplifier controls are grouped together, close to their input connectors; similarly, the sweep and triggering controls are conveniently located adjacent to each other.

Become familiar with the function of each control, know how to locate each control at a glance, and take the time to practice the procedure to display various waveforms on your scope. As with any tool, the better you know how and where to use it, the more valuable it will become.

To illustrate the initial setup and basic waveform display procedure, the B&K 1477P (panel and control layout shown in Fig. 2-7) dual-trace scope will be used. The setup procedure for other dual-trace scopes is similar.

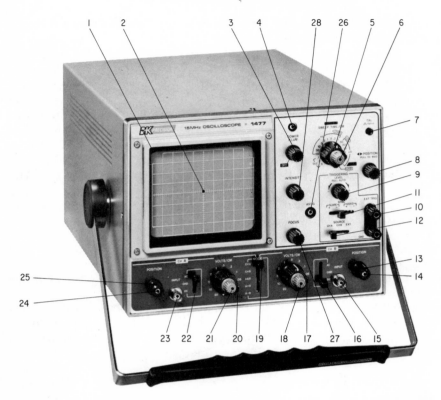

Fig. 2-7. Front panel controls of a dual-trace oscilloscope, B&K 1477P. *(Courtesy* Dynascan Corp.)

When operating a particular model scope for the first time, always take a few minutes to inspect the front panel controls. Make sure you are fully aware of what each control does. If in doubt, consult the instruction manual. A considerable amount of time can be wasted fiddling with controls when a few well-spent minutes reading the instruction manual can provide familiarity with proper operating procedures.

A dual-trace scope is generally purchased with the intent of waveform comparison with two signals displayed simultaneously. However, for many applications, only one signal need be observed. So first to be described are the procedure and applications for single-trace use; then dual-trace operation follows.

The setup and operating controls for a dual-trace triggered oscilloscope are basically similar to those of the single-trace model previously described. However, additional controls for triggering level, sync slope selector, mode switch for the dual-channel vertical amplifiers, etc. require explanation. A B&K 1477P, 15-MHz triggered scope is used in the following description; types supplied by other manufacturers are similar in control function and panel layout

Note how the functional layout of the controls (Fig. 2-7) makes it simple for the operator to quickly grasp the purpose of each adjustment. Both vertical channels, A and B, are below the CRT and are adjacent to each other. The sweep control and its associated trigger controls are grouped together for convenience. Finally, the CRT controls (focus, intensity, graticule illumination, and astigmatism) are located to the right of the CRT.

The location of the numbered controls is shown in Fig. 2-7.

Front Panel Layout of the B&K 1477P

1. Cathode Ray Tube (CRT)
2. The 8 × 10 cm graticule scale provides calibration marks for voltage (vertical) and time (horizontal) measurements. Illumination of the scale is fully adjustable.
3. POWER ILLUM control, ON-OFF control.
4. Pilot lamp.
5. SWEEP TIME/CM switch. Horizontal coarse sweep time selector. Selects calibrated sweep times of 0.5 μsec/cm to 0.5 sec/cm in 19 steps when VARIABLE control is set to the CAL position (fully clockwise). In the CH B position, this switch disables the internal sweep generator and permits the CH B input to provide horizontal sweep.
6. Sweep speed VARIABLE control. Fine sweep time adjustment. In the extreme clockwise (CAL) position, the sweep time is calibrated.
7. CAL IV P-P jack. Provides calibrated 1-kHz, 1-volt peak-to-peak square wave input signal for calibration of the vertical amplifier attenuators and to check the frequency compensation adjustment of the probes used with the oscilloscope.
8. ◀ ▶ POSITION control. Adjusts horizontal position of traces (both traces when operated in the dual-trace mode). Push-pull switch selects 5× magnification when pulled out (PULL 5× MAG); normal when pushed in.
9. TRIGGERING LEVEL control. Sync level adjustment determines points on waveform slope where sweep starts; (-) equals most negative point of triggering and (+) equals most positive point of triggering. Push-pull switch selects automatic triggering when pulled out (PULL AUTO). When there is automatic triggering, a sweep is generated even without an input signal.
10. EXT TRIG jack.
11. SYNC switch. Four-position lever switch with the following positions:
    SLOPE. The SLOPE positions are used for viewing all waveforms except television composite video signals.

(+) Sweep is triggered on positive-going slope of waveform. (−) Sweep is triggered on negative-going slope of waveform.

TV. In the TV positions, the sync pulses of a television composite video signal are used to trigger the sweep.

12. SOURCE switch. Three-position lever switch selects triggering source for the sweep. Both sweeps are triggered by the same source in dual-trace operation.

    CH A sweep is triggered by Channel A signal.

    CH B sweep is triggered by Channel B signal.

    EXT sweep is triggered by an external signal applied at the EXT SYNC jack.

13. Channel B POSITION control. Vertical position adjustment for Channel B trace. Becomes horizontal position adjustment when SWEEP TIME/CM switch is in the CH B position.
14. Channel B DC BAL adjustment.
15. Channel B INPUT jack.
16. Channel B DC-GND-AC switch.

    DC—Direct input of ac and dc component of input signal.

    GND—Opens signal path and grounds input to vertical amplifier. This provides a zero-signal base line, the position of which can be used as a reference when performing dc measurements.

    AC—Blocks dc component of input signal.

17. Channel B VOLTS/CM switch. Vertical sensitivity is calibrated in 11 steps from 0.01 to 20 V per cm when VARIABLE control is set to CAL position. This control adjusts horizontal sensitivity when the SWEEP TIME/CM switch is in the CH B position.
18. Channel B VARIABLE control. Vertical attenuator adjustment provides fine control of vertical sensitivity. In the extreme clockwise (CAL) position, the vertical attenuator is calibrated. This control becomes the fine horizontal gain control when the SWEEP TIME/CM switch is in the CH B position.
19. MODE switch. Five-position lever switch; selects the basic operating modes of the oscilloscope.

    CH A—Only the input signal to Channel A is displayed as a single trace.

    CH B—Only the input signal to Channel B is displayed as a single trace.

    A & B—Dual-trace operation; both the Channel A and Channel B input signals are displayed on two separate traces.

    A + B—The waveforms from Channel A and Channel B inputs are added and the sum is displayed as a single trace.

    A − B—The waveform from Channel B is subtracted from the Channel A waveform and the difference is displayed as a single trace. If only a Channel B input is present, the display is inverted.

20. Channel A VOLTS/CM switch. Same function as 17 except for Channel A.
21. Channel A VARIABLE control. Same function as 18.
22. Channel A DC-GND-AC switch. Same as 16.
23. Channel A INPUT jack.
24. Channel A DC BAL adjustment.
25. Channel A POSITION control.
26. ASTIG adjustment provides optimum spot roundness when used in conjunction with the FOCUS control and INTENSITY control.
27. FOCUS control.
28. INTENSITY control.

Oscilloscope Setup Procedure

1. Set POWER ILLUM control to OFF position (fully counterclockwise); refer to Fig. 2-7 for location of controls.
2. Connect power cord to a 120-V, 50/60-Hz outlet.
3. Set CH A POSITION control, CH B POSITION control 13, and ◄ ► POSITION to the centers of their ranges.
4. Pull TRIGGERING LEVEL control to the AUTO position.
5. Set CH A DC-GND-AC switch and CH B DC-GND AC switch to the GND positions.
6. Set the MODE switch to the CH B position for single-trace operation or the A & B position for dual-trace operation.
7. Turn on oscilloscope by rotating the POWER ILLUM control clockwise. It will "click" on and pilot lamp will light. Turn control clockwise to the desired scale illumination.
8. Wait a few seconds for the cathode ray tube (CRT) to warm up. A trace (two traces if operating in the A & B mode) should appear on the face of the CRT.
9. If no trace appears, increase (clockwise) the INTENSITY control setting until the trace is easily observed.
10. Adjust FOCUS control and INTENSITY control for the thinnest, sharpest trace.
11. Readjust position controls if necessary, to center the traces.
12. Check for proper adjustment of ASTIG control and DC BAL controls.

2.4 Single-Trace Waveform Observation on a Dual-Trace Scope

Either Channel A or Channel B can be used for single-trace operation. The advantage of using Channel B is that the polarity of the observed waveform can be reversed by placing the MODE switch in the A-B position if there is no input to Channel A. For convenience, Channel B is used in the following instructions.

1. Perform the steps of the "Initial Starting Procedure" with the MODE switch in the CH B position. Then connect the probe cable to the CH B INPUT jack. The following instructions assume the use of the B&K Precision Model PR-35 combination probes.
2. For all except low-amplitude waveforms, the probes are set for 10:1 attenuation. For low-amplitude waveforms (below 0.5 V peak-to-peak), set the probe for DIRect. The higher input impedance (low-capacity position) should be used when possible, to decrease circuit loading.
3. Set CH B DC-GND-AC switch to AC for measuring only the ac component (this is the normal position for most measurements, and must be used if the point being measured includes a large dc component). Use the DC position for measuring both the ac component and the dc reference, and any time a very low frequency waveform (below 5 Hz) is to be observed. The GND position is required only when a zero-signal ground reference is required, such as for dc voltage readings.
4. Connect ground clip of probe to chassis ground of the equipment under test. Connect the tip of the probe to the point in the circuit where the waveform is to be measured. If the equipment under test is a transformerless ac-powered item, use an isolation transformer to prevent dangerous electrical shock. The peak-to-peak voltage at the point of measurement should not exceed 600 V when the DIRect position of the probe is used.
5. Set CH B VOLTS/CM switch and the VARIABLE control to a position that gives 2 to 6 cm (2 to 6 large squares on the scale) vertical deflection. The display on the screen will probably be unsynchronized. The remaining steps are concerned with adjusting synchronization and sweep speed, which presents a stable display showing the desired number of waveforms. Any signal that produces at least 1 cm vertical deflection develops sufficient trigger signal to synchronize the sweep.
6. Set SOURCE switch to the CH B position. This provides internal sync so that the Channel B waveform being observed is also used to trigger the sweep. During single-trace operation on Channel A, the SOURCE switch should be placed in the CH A position for internal sync. Most waveforms should be viewed using internal sync; when an external sync source is required, the SOURCE switch should be placed in the EXT position and a cable should be connected from the EXT TRIG jack to the external sync source.
7. Set SYNC switch to the TV(+) or TV(-) positions for observing television composite video waveforms or to the SLOPE (+) or SLOPE (-) positions for observing all other types of waveforms. Use the (+)

position if the sweep is to be triggered by a positive-going wave, or the (−) position if the sweep is to be triggered by a negative-going wave. If the type of waveform is unknown, the SLOPE(+) position may be used.
8. Readjust TRIGGERING LEVEL control to obtain a synchronized display without jitter. As a starting point, the control may be pushed in and rotated to any point that will produce a sweep, which is usually somewhere in the center portion of its range. The trace will disappear if there is inadequate signal to trigger the sweep, such as when measuring DC or extremely low-amplitude waveforms. If no sweep can be obtained, pull the control out (PULL AUTO) for automatic triggering.
9. Set SWEEP TIME/CM switch and VARIABLE control for the desired number of waveforms. These controls may be set for viewing only a portion of a waveform, but the trace becomes progressively dimmer as a smaller portion is displayed. This is because the sweep speed increases but the sweep repetition rate does not change. When using very fast sweep speed at low repetition rates, you may wish to operate with the intensity control toward maximum.
10. After obtaining the desired number of waveforms, it is sometimes desirable to make a final adjustment of the TRIGGERING LEVEL control. The (−) direction selects the most negative point on the waveform at which sweep triggering will occur and the (+) direction selects the most positive point on the waveform at which sweep triggering will occur. The control may be adjusted to start the sweep on any desired portion of the waveform.
11. For a close-up view of a portion of the waveform, pull outward on the ◀ ▶ POSITION control. This expands the sweep by a factor of five (5× magnification) and displays only the center portion of the sweep. To view a portion to the left of the center, turn the ◀ ▶ POSITION control clockwise, and to view portions to the right of center, turn the control counterclockwise. Push inward on the control to return the sweep to the normal, nonmagnified condition.

2.5 Calibrated Voltage Measurements on a Dual-Trace Scope

Peak voltage, peak-to-peak voltages, dc voltages, and voltages of a specific portion of a complex waveform are easily and accurate measured, as shown in Fig. 2-8.

1. Adjust controls as previously instructed to display the waveform to be measured.
2. Be sure the CH B vertical VARIABLE control is set fully clockwise to the CAL position.
3. Set CH B VOLTS/CM switch for the maximum vertical deflection possible without exceeding the limits of the vertical scale.

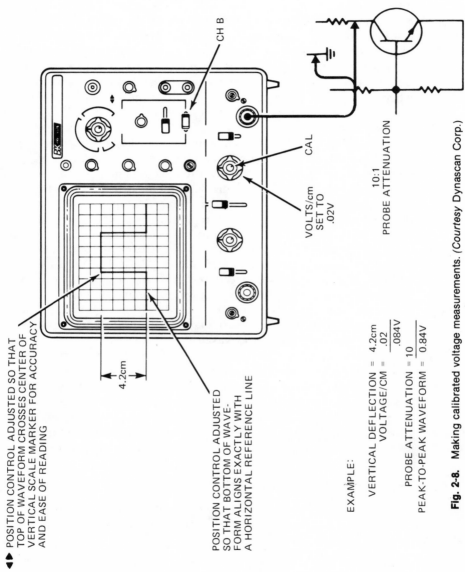

Fig. 2-8. Making calibrated voltage measurements. *(Courtesy Dynascan Corp.)*

4. Read the amount of vertical deflection (in cm) from the scale. The CH B POSITION control may be readjusted to shift the reference point for easier scale reading if desired. When measuring a dc voltage, adjust the CH B POSITION control to a convenient reference with the CH B DC-GND-AC switch in the GND position, then note the amount the trace is deflected when the switch is placed in the dc position. The trace deflects upward for a positive voltage input and downward for a negative voltage input. For an accurate display of high-frequency waveforms above 10 MHz, it is important that (1) the probe be used in the 10:1 position to reduce circuit loading; (2) the oscilloscope controls be set so that the height of the pattern does not exceed 4 cm; and (3) the trace be centered vertically.
5. Calculate the voltage reading as follows: Multiply the vertical deflection (in cm) by the VOLTS/CM control setting (see example in Fig. 2-8). Don't forget that the voltage reading displayed on the oscilloscope is only 1/10th the actual voltage being measured when the probe is set for 10:1 attenuation. The actual voltage is displayed only when the probe is set for DIRect measurement.
6. Calibration accuracy may be occasionally checked by observing the 1-V peak-to-peak square wave signal available at the CAL 1V P-P jack. This calibrated source should read exactly 1 V peak-to-peak.

2.6 Operating Precautions

Certain operating precautions apply to the operation of *any* oscilloscope:

1. Study the manufacturer's instruction manual thoroughly before attempting to use the oscilloscope. Do this even if you have had experience with other oscilloscopes.
2. Always place the instrument into operation according to the steps given. Never connect the instrument to the test circuit until these steps are completed.
3. Switch the test-circuit power off until all connections to the oscilloscope are completed.
4. Use the minimum intensity necessary for comfortable viewing or efficient photographing. Do not operate the oscilloscope in bright room lighting or sunlight.
5. Keep the spot moving on the screen. When it must stand still, reduce intensity to keep the screen from burning.
6. Unless the CRT is flat-faced, make all measurements in the center area, or reading errors will be caused by distortion in the curved periphery.
7. Always connect the GROUND (COMMON) terminal of the oscilloscope to the ground, or low-potential, point in the test circuit. Do this *before* connecting the high input terminal. Disconnect the instrument in the opposite sequence.

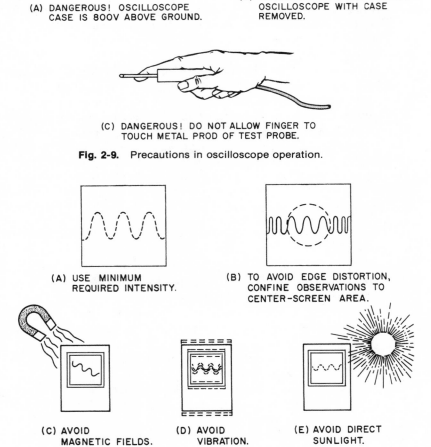

Fig. 2-9. Precautions in oscilloscope operation.

Fig. 2-10. Hints for oscilloscope operation.

CAUTION: *When both points in the test circuit are above ground, the metal case of the oscilloscope will be connected to high voltage and could deliver a dangerous electric shock.*

8. Use only shielded probes, and keep the fingers well away from the metal prod or tip.
9. Your oscilloscope is shielded; nevertheless, you should keep it clear of strong magnetic fields. Such fields can distort the display.
10. Never operate the instrument at a higher line voltage or different power frequency than that recommended by the manufacturer.
11. Keep all input-signal voltages below the specified maximum.
12. Never operate the instrument outside of its case without being extra careful. Many circuit points which thus would be exposed carry dangerously high voltage. Another hazard is the CRT, which can implode and scatter glass with great force.
13. Observe all of the well-known electrical safety rules when working with an oscilloscope (Fig. 2-9).
14. Protect the instrument from vibration and mechanical shock (Fig. 2-10).
15. Be sure that any internal fan or blower is operating.
16. Clean ventilating air filters regularly.
17. Replace the CRT when a bright trace no longer can be obtained or when spots have been burned into the face. Replace small tubes or transistors and/or make repairs as soon as degraded or intermittent operation is noticed. Refer to Chapter 7 for service advice.
18. Leave repairs and adjustments to a skilled instrument technician. If you must adjust the operating controls, first study the oscilloscope manufacturer's instructions carefully.
19. The oscilloscope is not a plaything. Neither participate in nor condone pranks and horseplay with this instrument.

3

oscilloscope accessories

For most applications, the oscilloscope is used without external devices. In other uses, some attachment, such as a probe, is needed to modify the response of the instrument in some desired way or to modify the input signal. The operator must be familiar with these accessories if he is to use them profitably.

3.1 Why Special Probes Are Needed

From experience with test meters, the technician knows that a probe functions simply as an extension of the meter terminal. Basically it is nothing more than a slender metallic prod on the end of an insulated rod (handle) connected to the instrument terminal through a flexible insulated lead. Through its use, the instrument can be connected to any circuit point which the prod touches. This is a probe in its simplest form.

The simple probe may be used to connect an oscilloscope quickly to any desired point in a test circuit. This is entirely satisfactory at dc and low audio frequencies and when the maximum amplification of the oscilloscope channels is not used. At high frequencies and/or high gain, however, the simple probe is inadequate: hand capacitance can cause hum pickup or attentuation of the signal amplitude, and the input impedance of the oscilloscope (which the probe transfers directly to the test point) may be low enough to load the test circuit. To minimize hand capacitance and stray coupling, the probe handle and lead must be carefully shielded; the capacitance between shielding and conductor must be minimal. To reduce loading effect by increasing the impedance of the instrument circuit, a

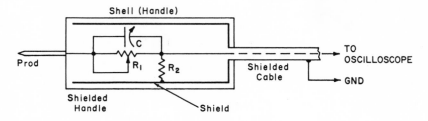

Fig. 3-1. A low-capacitance probe.

suitable series capacitor is built into the shielded probe (sometimes a series resistor is used).

Other special probes convert the signal to a form that may be handled and displayed by the oscilloscope. Thus, a *voltage-divider probe* reduces signal voltage to a safe level; a *demodulator probe* selects the modulating voltage from an AM wave; and an *RF probe* rectifies the radio-frequency voltage and delivers a dc output proportional to the peak RF voltage.

3.2 Low-Capacitance Probe

The low-capacitance probe decreases the input capacitance of the measuring circuit. Its use ensures that the circuit under test will be least disturbed by connection of the instrument. For example, without this probe, connection of the oscilloscope can completely detune a radio-frequency circuit.

Figure 3-1 shows the details of a low-capacitance probe. Resistors R_1 and R_2 and trimmer capacitor C are contained in the shielded handle and are preset by screwdriver adjustment. The resistances generally are selected so that, with the capacitance, they form a 10:1 voltage divider when the probe is connected to the oscilloscope vertical amplifier input. Capacitor C provides frequency compensation to preserve the frequency response of the oscilloscope. C and R_1 are factory-adjusted and need not be disturbed unless the probe is being recalibrated.

The values of C, R_1, and R_2 are chosen, with respect to the input impedance of the oscilloscope and the shunt capacitance of the shielded cable, so that the 10:1 voltage division and a 4:1 to 10:1 capacitance reduction are obtained. The operator using a voltage-calibrated oscilloscope must remember that voltage indications will be one-tenth of true values when the probe is attached.

Some low-capacitance probes are provided with a switch which at one setting short-circuits C and R_1 to give direct input to the oscilloscope. This permits the single probe to be used both for low-capacitance and straight-through input.

3.3 Resistor-Type Voltage-Divider Probe

The maximum signal voltage that can be handled safely by the input channels of an oscilloscope depends on the ratings of components and insulation in these channels. When higher voltages are to be checked, they must be reduced to a safe value by means of a voltage-divider probe (also called *high-voltage probe*).

One such probe employs a simple resistance voltage divider, as shown in Fig. 3-2. Resistors R_1 and R_2 are chosen in value, with respect to the input impedance of the oscilloscope vertical amplifier, so that the desired voltage stepdown will be obtained (this usually is 10:1 or 100:1). The variable resistance R_1 may be set initially for exact division.

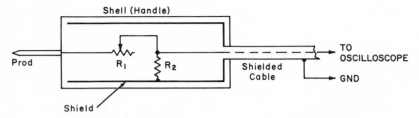

Fig. 3-2. A resistor-type, voltage-divider probe.

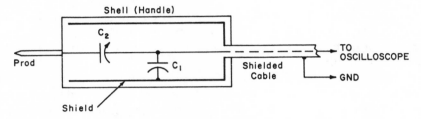

Fig. 3-3. A capacitor-type, voltage-divider probe.

The low-capacitance probe (Fig. 3-1) may also be used, provided that its voltage division ratio is adequate. In addition to signal reduction, the low-capacitance probe provides flat frequency response because of its frequency compensation.

3.4 Capacitor-Type Voltage-Divider Probe

The resistor-type probe just described is unsuitable in some applications in which the probe resistors set up a stray conduction path in the circuit under test. In such applications, a probe may be employed which contains a capacitive voltage divider.

In the probe shown in Fig. 3-3, voltage division is effected by capacitors C_1 and C_2. These capacitances are chosen, with respect to the shunting capacitance of the cable and the oscilloscope input capacitance, to give the required voltage division. C_2 is made variable for initial adjustment of the output voltage to the exact desired value.

3.5 Demodulator Probe

In order to display the low-frequency component (modulation envelope) of an amplitude-modulated signal, the signal must be demodulated before it is presented to the oscilloscope. This is accomplished with a demodulator probe (Fig. 3-4), which essentially is a diode detector.

In this probe, C is a low-capacitance dc-blocking capacitor which protects the germanium diode (D) from damage by any dc component in

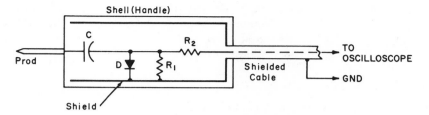

Fig. 3-4. A demodulator probe.

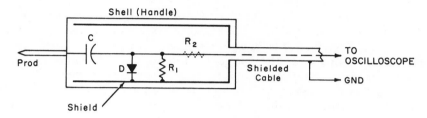

Fig. 3-5. An RF probe.

the circuit under test. A shunt rectifier circuit is formed by C, D, and R_1. As a result of the action of this circuit, the diode detects (demodulates) the applied AM signal and develops a voltage across load resistor R_1 that has the waveform and frequency of the modulating voltage and is proportional to its amplitude. This action is the same as that of a diode radio detector which demodulates the radio signal and delivers the audio component. R_2 is an isolating resistor which minimizes loading effect of the probe and oscilloscope on the circuit under test.

3.6 RF Probe

When the oscilloscope response is short of the RF spectrum, radio-frequency voltages may be measured only if an RF probe is attached. This probe rectifies the RF energy and delivers a dc output voltage which is almost equal to the peak RF voltage. This dc voltage is then applied to the oscilloscope dc vertical input and measured on the calibrated viewing screen.

Figure 3-5 shows the structure of a typical RF probe, which is similar in circuit to the modulator probe (Fig. 3-4). Here, however, blocking capacitor C has a much higher capacitance than the one used in the demodulator probe. It is rated between 0.02 and 0.05 μF to ensure peak-voltage operation of the shunt-diode rectifier circuit. Germanium diode D rectifies the RF voltage and develops a dc output voltage across load resistor R_1 equal to the peak RF voltage minus the small forward drop across the diode. The resulting oscilloscope deflection is read as peak RF voltage. When rms indications are desired, a series resistor (R_2) is included in the probe to drop the output voltage by the proper amount. This resistance is chosen with respect to the oscilloscope input resistance, so that a voltage division of 0.707 is provided.

3.7 Dc Voltage Calibrator

When the oscilloscope is used for voltage measurements, provision must be made for voltage-calibrating the screen. Some oscilloscopes supply an internally generated voltage for this purpose. A service-type instrument, for example, may supply a 1-V peak-to-peak, line-frequency, sine-wave potential at a front-panel terminal. For calibration purposes, this voltage is applied to the input channel (usually, the vertical) and the gain control is adjusted to align the top and bottom of the pattern with calibration points on the graticule. A professional, laboratory-type oscilloscope might supply an internally generated 400- or 1000-Hz square-wave voltage through an attenuator (coarse and fine) which reads directly in volts and millivolts. This adjustable calibrating voltage is applied to the vertical channel whenever the selector switch of the channel is set to its CALIBRATE position.

When an oscilloscope supplies no internal calibrating voltage, or one of unsuitable amplitude, an external source must be used. Both ac and dc continuously variable voltage calibrators are available for this purpose.

Figure 3-6 shows the details of a dc voltage calibrator. This unit contains a closely regulated 100-V dc supply and an attenuator consisting of potentiometer R_1 and a range selector (switch S and resistors R_2 to R_5). R_1 is direct-reading in voltage (0-100), and its indications are multiplied by 0.001, 0.01, 0.1, or 1, depending on the setting of the switch. The output voltage indicated by the combined readings of the potentiometer and

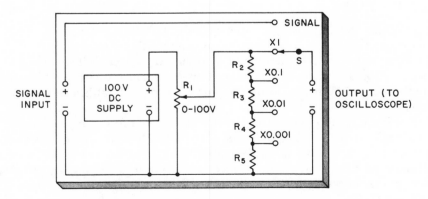

Fig. 3-6. A dc voltage calibrator.

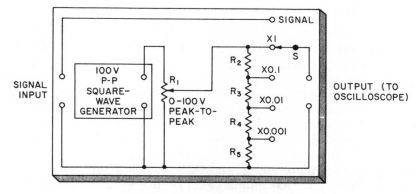

Fig. 3-7. An ac voltage calibrator.

switch is the voltage presented to the oscilloscope. When S is set to its SIGNAL position, the signal under observation (applied to the SIGNAL INPUT terminals) is transmitted through the calibrator, minus the calibrating voltage, to the oscilloscope. This arrangement enables the operator to place on the screen, at will, the signal or calibrating voltage for calibration.

3.8 Ac Voltage Calibrator

An ac voltage calibrator is similar to the dc calibrator just described, except that the internal source supplies a voltage-regulated, line-frequency, square-wave signal at 100 V peak-to-peak. The fine-control potentiometer (R_1 in Fig. 3-7) is direct-reading 0-100 V p-p, and this voltage is multiplied by the setting of switch S (0.001, 0.01, 0.1, or 1). The indicated value of peak-to-peak voltage is applied to the oscilloscope.

As with the dc calibrator, when switch S is set to SIGNAL, the signal under study (applied to the SIGNAL INPUT terminals) is transmitted through the calibrator, minus the calibrating voltage, to the oscilloscope. This arrangement enables the operator to place on the screen, at will, the signal or the calibrating voltage for comparison.

3.9 Frequency (Time) Calibrator

Some oscilloscopes supply an internally generated standard frequency for calibrating the sweep or time axis. This is a sine-wave or square-wave voltage of accurate frequency. The calibrating voltage may be applied to the vertical input, and the sweep frequency and sync adjusted for a single stationary cycle on the screen.

Common frequencies are 1000 Hz (1 cycle = 1 millisecond), 100 kHz (1 cycle = 10 microseconds), and 1 MHz (1 cycle = 1 microsecond).

3.10 Electronic Switch

The *electronic switch* is a device that enables two signals to be displayed simultaneously on the screen of a single-gun CRT. (Electronic switches have also been designed for more than two displays.) This performance is a convenience, since it permits the direct comparison of two signals without necessitating a multiple-gun tube.

Some advanced professional oscilloscopes have self-contained electronic switches. But less costly instruments do not offer this feature, and an external electronic switch must be used with them.

Figure 3-8 shows the skeleton circuit of an electronic switch. Here, Q_1 and Q_2 are amplifiers, and Q_3 and Q_4 are switches. Input signal No. 1 is applied to amplifier Q_1 through gain control R_1; input signal No. 2 is applied to Q_2 through R_2. The square-wave generator alternately biases first Q_3 and then Q_4 to cutoff. When Q_3 is cut off, Q_4 is conducting and transmits input signal No. 2 to the OUTPUT terminals. Conversely, when Q_4 is cut off, Q_3 is conducting and transmits input signal No. 1 to the OUTPUT terminals. When the square-wave switching frequency is much higher than either signal frequency, bits of each signal are alternately presented to the oscilloscope vertical input to reproduce the two signals on the screen.

Figure 3-9 shows some typical two-signal displays afforded by the electronic switch. The traces are composed of tiny segments representing the switching intervals, but when the switching rate is fast enough, the segments are so small that a solid line is seen. In Fig. 3-9A, the traces are laid on top of each other for easy comparison by adjustment of the POSITION CONTROL potentiometer (R_5 in Fig. 3-8). In Fig. 3-9B and C, the traces have been separated by readjustment of R_5. The traces may be

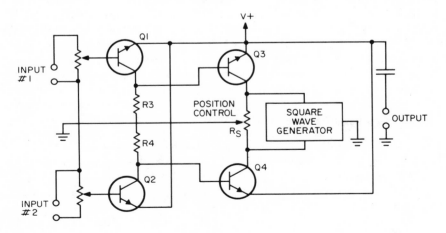

Fig. 3-8. An electronic switch.

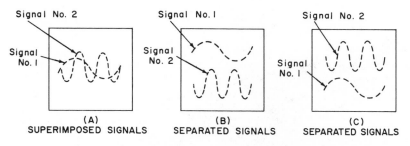

Fig. 3-9. Oscilloscope display through an electronic switch.

shifted vertically with respect to each other to any desired position for easy comparison of amplitude, phase, and frequency, and their individual heights may be adjusted by means of gain controls R_1 and R_2.

3.11 External Amplifiers

Although high-gain amplifiers are employed in an oscilloscope, some signals are too weak to produce a large enough pattern. In order to accommodate these signals, external amplification must be provided.

An auxiliary amplifier must supply the required extra gain and must have frequency response identical with that of the oscilloscope channel into which it feeds. The shortest possible leads should be used to connect it to the oscilloscope, and these should be shielded if practicable.

Conventional external amplifiers of the wideband type supply voltage gains of 20 dB and 40 dB that can be selected with a switch.

3.12 Trace Photography

There are two general ways to photograph displays on a CRT. One is to stock up on a lot of film and take numerous photos while varying shutter speed and lens openings; with a bit of luck, some of the final photos will be acceptable, and expensive.

A second, and far less expensive approach, is to work within a narrow limit of CRT/camera/film combinations and produce fine trace photographs. The key parameters involved in waveform photography include the CRT trace intensity, phosphor type, graticule illumination, and scan time; the camera adds lens opening, film speed, and shutter opening. These seven parameters combine to determine writing speed.

Trace photography is the technique to expose the camera's film to the exact amount of light emitted from the CRT screen without overexposure (too bright and washed out) or too dim (trace difficult to analyze).

The trace, the graticule illumination, and the CRT background contribute to the light output and contrast. Unfortunately, it is not as simple as looking through the camera viewfinder and letting the eye judge if conditions are optimum. The eye has a very broad dynamic range and

Fig. 3-10. The recording camera assembly for an oscilloscope. A solid-state electronically controlled shutter is included in this scope camera. *(Courtesy* Hewlett-Packard.)

can sense light intensity over a range of one hundred or so; the eye's response is logarithmic while the film's is linear. Second, the eye and film have a different sensitivity to color. The most common phosphor used on CRTs is P31, peaked at 530 nanometers or greenish, which is good for human operator viewing. However, a blue or violet trace, with a wavelength of 450 nanometers, is preferable for trace photography since most films peak near this region. For this reason, panchromatic film is generally used since it has a fairly uniform response over the entire visible spectrum. "Instant" cameras, such as offered by Polaroid, Kodak, and several others, are convenient since immediate results can be seen.

To take photographs of a CRT display, first set the scope on a solid table or bench and the camera on a sturdy tripod or scope camera attachment; avoid hand-held shots. Adjust the trace intensity for adequate light output, taking care to avoid blooming or halo effects due to excessive phosphor bombardment. Next adjust the graticule illumination to the point where the vertical and horizontal lines are clear but not dominant. Carefully check the focus and astigmatism settings so that the beam is sharp. Do all viewing through the camera viewfinder rather than by trying to judge trace intensity by eye in a well-lit room.

Estimate the sweep trace time and set the shutter speed to include at least four sweeps. Use the Table of Recommended Films and select an f-stop (lens opening). Assume f8 with one second is the first attempt. Take a picture, wait for it to develop, and closely examine the results. If it is too light or overexposed, setting the lens opening one stop higher (or f11) will allow less light through the lens. If the trace is too low in intensity or underexposed, use a lower f stop (f5.6) and more light will be allowed to strike the film.

If trace afterglow is apparent, turn down the intensity control and allow a longer exposure time. If jitter is apparent on the photograph, the only solution may be to readjust the camera shutter speed so that two, rather than four, sweeps occur while the shutter is open. Remember, if the shutter speed is reduced say from one second to a half second, the lens opening must be reduced one stop (from f8 to f5.6) to achieve the same previous shutter speed/lens opening relationship. Now the camera shutter is open for half the time, but the lens opening is also twice as wide.

A modern oscilloscope recorder consists of a Polaroid Land[1] camera equipped with a special lens and a mounting frame that fastens to the front of the oscilloscope and provides a hood system for viewing the screen. Because the Polaroid camera gives a finished picture in a few seconds without darkroom processing, records are quickly available at each step in a test.

Figure 3-10 shows a camera assembly of this type.[2] The unit fastens snugly to the front of the oscilloscope. The familiar Polaroid camera back is at the lower left, and the viewing hood extends diagonally upward above it. The mounting adaptor is hinged so that the entire camera assembly may be swung horizontally away from the oscilloscope when direct access is desired.

Characteristics. Oscilloscope cameras are available with a number of different features. The model shown in Fig. 3-10, for example, normally is prefocused but has an adjustment for sharp focus. The camera back may be moved horizontally or vertically, in relation to the lens, through five detented positions for recording up to five separate displays on one frame. The back may also be rotated through 90° increments. While the standard Polaroid roll-film back is shown, and is used most often (giving 3¼" × 4¼" prints), the assembly will also accept backs for 4" × 5" Polaroid sheet film, conventional cut film from 2¼" × 3¼" to 4" × 5" or a film-pack. Oscilloscope camera manufacturers provide a good selection of lens and shutter characteristics.

[1]Registered trademark of Polaroid Corporation, Cambridge, Mass.
[2]Type 197B Hewlett-Packard.

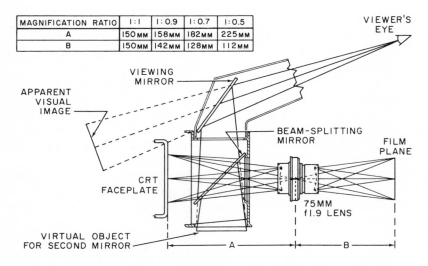

Fig. 3-11. The optical system of a typical CRT camera. (*Courtesy* Tektronix.)

Optical Arrangement. Figure 3-11 shows the arrangement of the camera assembly. (The following explanation of its operation is simplified.) Rays of light from the trace and the illuminated graticule lines in the oscilloscope enter the system. In the path of these rays is the *beam-splitting mirror*. This is a transparent mirror mounted at a 45° angle. It transmits some of the rays through to the camera (represented here by lens and film plane) and reflects other rays up to the *viewing mirror*. The operator looks into the viewing mirror and sees the oscilloscope display as a virtual image. With this optical system the display can be viewed with the camera in place, even while the picture is being taken.

The term *writing rate* in oscilloscope photography designates the highest spot speed which will produce an acceptable picture. It is expressed in centimeters per microsecond. High writing rate is needed to record fast transients, whereas a low writing rate will satisfactorily reproduce recurrent phenomena or slow sweeps. (The CRT phosphor influences the writing rate. Arranged in decreasing order starting with the highest writing rate, CRT phosphors commonly found in oscilloscope tubes are P11, P2, P1, and P7.) Film must be selected with respect to writing rate.

Typical Operation. The selection of film type, exposure time, and type of display depends ultimately on the type of phenomenon to be recorded and CRT characteristics. With a given lens aperature, the shutter speed, for example, will be fast for a bright stationary or repetitive display and slow for a dim display. When recording fast transients and other high-speed, single-shot phenomena, the shutter should be held open (bulb position) while a single sweep is triggered manually, and then closed.

A dimmer, thin trace will give the best recording, but requires longer exposure. A very bright trace tends to produce halo or afterglow. The oscilloscope screen illumination must brighten the graticule lines sufficiently for sharp, clear reproduction but must not produce glare. If the display is viewed during exposure, the operator must keep his face against the hood to prevent entry of light.

The ability to make multiple exposures on a single frame is advantageous in many tests and measurements where either separate signals or the same signal at different stages must be shown close to each other for comparison. For this purpose, the camera may be slid and locked into as many as five successive vertical positions for as many separate exposures on the same frame.

General steps in making oscillogram records with a self-processing oscilloscope camera are as follows:

1. Load camera with film having proper writing rate for type of test.
2. Set up oscilloscope and make dry run of test to ensure correct operation of equipment.
3. Attach camera to oscilloscope.
4. Test-operate oscilloscope while observing screen through viewing hood of camera assembly.
5. Determine exposure to be used.
6. Photograph display, following any special directions given by manufacturer of camera or oscilloscope.

The use of conventional film for recording from the oscilloscope has the disadvantage that darkroom processing is required. But it is used when (1) a self-processing camera is not available; (2) economy is a factor; (3) a continuous film strip is desired; or (4) processing delay is of no concern. The time between taking and viewing a picture may be reduced by accepting the developed negative in place of the paper print.

Some oscilloscope camera assemblies will accept special conventional-camera backs. In the absence of a special oscilloscope camera or adaptor, any good camera may be used if it is properly focused on the oscilloscope screen, loaded with satisfactory film, and correctly adjusted for lens opening and shutter speed.

Conventional film, like Polaroid film, must be selected with due regard to writing rate. A fast film, such as one of the panchromatic types, is needed for rapid transients and other single-shot phenomena, whereas a slow film, such as one of the orthochromatic types, will serve for slow, stationary, or recurrent phenomena. With a given film, long exposure is required for a satisfactory picture of a dim, slow-speed trace; short exposure for a bright or fast trace. Aperture size and shutter speed depend upon trace brightness and type of film.

There are two requisites for successful use of a conventional camera: (1) film-plane focusing (this demands a ground glass-back) and (2) a hood

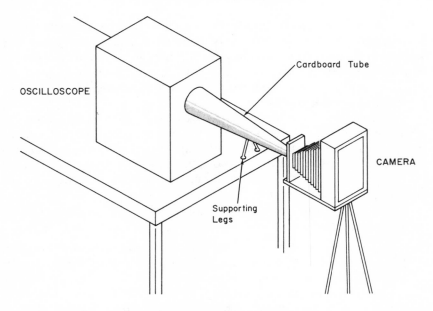

Fig. 3-12. A conventional camera setup.

of some kind to exclude ambient light. The latter is necessary because a camera with the usual lens must be moved farther back from the screen than an oscilloscope camera need be.

A simple hood consists of a cone-shaped cardboard tube run between the camera and oscilloscope screen and fitted snugly at each end to prevent light leaks (Fig. 3-12). The inside of the tube must be painted dull black to kill reflections. For protection against movement, the camera must be rigidly supported either with a tripod, as shown in Fig. 3-12, or by resting it on the table with the oscilloscope. The required length of the tube depends upon the focal length of the lens. To determine proper length experimentally, set up the camera in a dimly lighted room and focus it sharply on the oscilloscope trace. Then, measure the distance from screen to lens and cut the cone to that length. (A camera such as the small 35 mm models, which have no ground glass, may be opened and a piece of waxed paper stretched and taped temporarily across its open back to substitute for ground glass during focusing.) In a completely darkened room, a hood will not be needed.

The following are general steps in recording oscillograms with a conventional camera prefocused as explained above:

1. Load camera with film having proper speed for type of test.
2. Set up oscilloscope and make dry run of test to ensure correct operation of equipment.

3. Set up camera with light hood.
4. Determine exposure time and lens opening to be used, and set lens and shutter accordingly.
5. Photograph display.
6. Develop film.

For good recording of the lines, the graticule lighting should be turned up high enough to make the lines shine but not so high as to produce glare. Sometimes, especially when exposure is short, this will not give sharp reproduction.

Improvement is afforded by double exposure. Expose the film first with the graticule illuminated but no trace on the screen, then expose it a second time for the trace. The graticule exposure time cannot be prescribed exactly; it should be found experimentally with each type of film used and recorded for future reference.

In a motion-picture camera, the film is advanced one frame at a time, each frame pausing behind the lens during exposure. In the *moving-film* camera used for special oscilloscope photography, the film is drawn past the lens continuously by an adjustable-speed motor. There is no blinking shutter. This action permits recording of a phenomenon which starts at some unpredictable instant (e.g., an occasional surge or a random pulse).

The film movement provides the horizontal sweep and, therefore, the time base. For this reason, the oscilloscope sweep must be switched off and only vertical spot deflection used. The developed film will still show the complete oscillogram referred to both axes, for the sweep is supplied mechanically by the camera. When a camera is used in which the fim travels vertically with respect to the oscilloscope, the signal must be applied to the horizontal input of the oscilloscope, with the vertical input switched off. If the horizontal channel passband is too narrow for a projected test, the signal must be applied to the vertical channel (as in the first example) and either the camera or the oscilloscope laid on its side.

In the absence of recording equipment, a hand tracing may be made from an oscilloscope screen. This requires more care than skill and is limited to stationary patterns or those of long duration.

A disk of transparent plastic is best for hand recording, as it gives a clear view of the trace. It must be thick enough to prevent parallax but not so thin that it puckers or wrinkles—0.01 inch is satisfactory. A graticule may be inscribed on one face and the tracing done on the opposite one.

To make a record:

1. Place the disk over the face of the oscilloscope.
2. Carefully align the graticule of the disk with that of the oscilloscope.
3. Either attach the disk to the face with two or three small dots of two-sided adhesive tape or hold it firmly in place.
4. Use a well-sharpened grease pencil to trace the pattern. This pencil gives a solid, readable line which later is easily removed by rubbing with tissue. It also does not scratch the plastic.

A thin paper disk may be used in place of the transparent plastic but it requires a brighter trace. The disk may be cut from draftsman's tracing paper, and a graticule rule on one side.

The recording procedure is the same as that for the transparent disk, except that a medium pencil (grade HB lead) should be used.

Discharging Electrified Disk. A plastic or paper disk may become charged with static electricity when it is lifted from its storage place, especially if it is drawn across a dry, polished surface. When it is placed against the CRT face, its charge may distort the trace. To prevent this, discharge the disk before attaching it to the oscilloscope by touching it momentarily to a cold-water pipe or by breathing on it.

4

voltage and current measurement

A conventional CRT with calibrated screen is fundamentally an electrostatic voltmeter. The oscilloscope, therefore, may be used directly as a voltmeter and indirectly as a current meter. But, unlike other electronic meters, the oscilloscope can show waveform, frequency, and phase, as well as amplitude of a current or voltage. Still another advantage is the exceedingly fast response of the oscilloscope.

In spite of its utility, however, the oscilloscope is not a precision voltmeter; in much work of a routine or practical nature, it certainly could not be used economically in place of a simpler and less expensive meter. It is important because of its wide frequency response, lack of inertia, and ability to display waveform, and because it can indicate voltage in addition to, and simultaneously with, other phenomena.

4.1 How to Voltage-Calibrate the Screen

Before an oscilloscope can be used for direct measurement of voltage or current, its screen must be calibrated. The calibration procedure depends upon the type of oscilloscope and whether or not an external calibrating source must be used. Recommended procedures are given below. In each example, the instruction "set up the oscilloscope" means to place the instrument into operation.

Calibration Procedure—with Internal Single-Voltage Ac Source

1. Set up oscilloscope.
2. Connect jumper between CALIBRATION VOLTAGE terminal and VERTICAL INPUT terminal (Fig. 4-1A).
3. Advance VERTICAL GAIN control for readable pattern.
4. Set SYNC selector to INTERNAL.
5. Adjust sweep frequency and SYNC control for several stationary cycles (Fig. 4-1B).
6. Adjust HORIZONTAL GAIN control to spread this pattern, as desired, on screen.

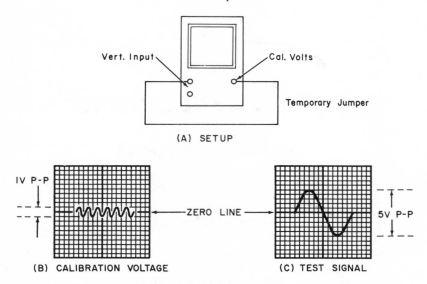

Fig. 4-1. Calibration with internal single voltage.

7. Adjust VERTICAL GAIN control to align tips of positive half-cycles and tips of negative half-cycles with corresponding marked calibration lines on screen.

The instrument is now calibrated. The calibration voltage usually is 1 V peak-to-peak, so the vertical distance between the points represents 1 V p-p. The VERTICAL ATTENUATOR, if present, will multiply indications in this range by 0.1, 1, 10, etc., depending upon the setting.

The manufacturer's instruction manual generally specifies the setting of the VERTICAL GAIN control (or VERTICAL ATTENUATOR) for calibration. The gain control always must be returned to the calibration setting when direct voltage measurements are to be made.

Some oscilloscopes do not bring the calibrating voltage to a panel terminal. Instead, this voltage is transmitted internally to the vertical channel when the VERTICAL SELECTOR switch is set to its CALIBRATE position.

The VERTICAL GAIN control may also be set to spread the calibration voltage over any desired vertical length. Thus, if this control is set for deflection of one division above and below the zero line for 1 V p-p, as shown in Fig. 4-1B, each scale division *at that setting of the gain control* represents 0.5 V p-p. If a subsequently applied test signal occupies 10 divisions, as shown in Fig. 4-1C, its voltage then is read as 5 V p-p. Some oscilloscopes have a graticule reading directly in volts, these indications being multiplied by settings of the vertical attenuator.

With the internal sweep switched off, a single vertical line is obtained instead of the ac cycles. The length of the line is proportional to the voltage.

Professional, laboratory-type oscilloscopes provide an internally generated, continuously variable, square-wave calibrating voltage. No external connections are required. The following procedure should be followed to calibrate the screen with this arrangement:

Calibration Procedure—with Internal Variable-Voltage Ac Source

1. Set up oscilloscope.
2. Set **VERTICAL SELECTOR** to **CALIBRATE**.
3. Switch on calibrator.
4. Set internal sweep to lower frequency than that of calibrator.
5. Set **VERTICAL GAIN** control to desired operating position.
6. Adjust **CALIBRATOR** control to align flat peaks of square wave with the desired vertical scale divisions (see Fig. 4-2A).
7. Adjust **SYNC** control for stationary pattern, with **SYNC SELECTOR** set to **INTERNAL**.
8. Read peak-to-peak voltage from scale of **CALIBRATOR** control.

In Fig. 4-2, the calibrating voltage has been adjusted for deflection of two divisions above and below the zero line. If the voltage is read from the CALIBRATION control as 4 V p-p, the resulting screen calibration figure is 1 V p-p/div. (The full vertical axis shown in the illustration—20 divisions—thus would represent 20 V p-p.)

When the sweep frequency (f_s) is set to equal the frequency of the calibrating voltage (f_c), a single square-wave cycle will be seen; when f_s is lower than f_c, several cycles will appear (as in Fig. 4-2A); and when f_s is much lower than f_c, the vertical lines of the pattern will become obscured and the flat peaks will merge to form two horizontal lines (as in Fig. 4-2B).

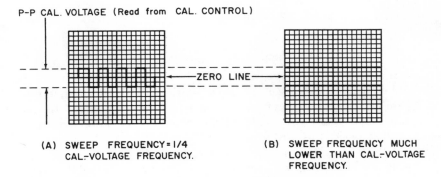

Fig. 4-2. Internal-calibrator patterns.

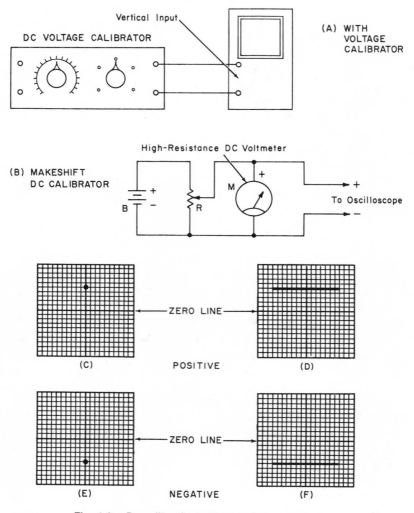

Fig. 4-3. Dc calibration with an external source.

If the internal sweep is switched off, a single vertical line is obtained, its length indicating the voltage.

When the square wave is applied to the dc vertical amplifier, its peak above the zero line indicates positive voltage, and its peak below the zero line indicates negative voltage. Each of these voltages is equal to the *peak ac* (½ peak-to-peak) indicated by the **CALIBRATOR** control. Thus, Fig. 4-2 would show a deflection of +2 V and -2 V dc.

When the oscilloscope provides no internal calibrating voltage, an external source must be used. Voltage calibrators suitable for this purpose are described in Sections 3.7 and 3.8.

Calibration Procedure—with External Source
For dc voltage calibration:
1. Set up dc oscilloscope.
2. Switch off internal sweep.
3. Connect dc calibrator, as shown in Fig. 4-3A.
4. Set calibrator controls for desired calibrating voltage.
5. Adjust VERTICAL GAIN control to move spot vertically up to the desired number of screen divisions. Thus, the five-division deflection in Fig. 4-3C indicates a positive voltage of the value shown by settings of the calibrator controls.
6. Reverse the leads to the oscilloscope, noting that the same deflection occurs vertically downward, as in Fig. 4-3E. This indicates a negative voltage of the value shown by settings of the calibrator controls.
7. When the internal sweep is switched on, a horizontal line trace is obtained, and this line moves above or below the zero line to indicate a positive voltage (Fig. 4-3D) or negative voltage (Fig. 4-3F), respectively.

If a calibrator is not available, a battery, potentiometer, and high-resistance voltmeter may be used (Fig. 4-3B), the calibrating voltage being adjusted with potentiometer R and read from meter M.

For ac voltage calibration:
1. Set up oscilloscope.
2. Switch on internal sweep.
3. Connect ac calibrator, as shown in Fig. 4-4A.
4. Set calibrator controls for desired calibrating voltage.
5. Adjust VERTICAL GAIN control to spread square-wave pattern vertically between desired screen divisions. The pattern will resemble Fig. 4-2A or B, depending on sweep frequency. This deflection corresponds to voltage indicated by settings of calibrator controls.

If a calibrator is not available, a transformer, potentiometer, and high-impedance voltmeter may be used (Fig. 4-4B), the calibrating voltage being adjusted with potentiometer R and read from meter M.

4.2 Direct Measurement of Voltage

In direct measurement of voltage, a calibrated oscilloscope is used like a voltmeter. The instrument must previously have been voltage-calibrated following one of the methods outlined in Section 4.1. Use short leads between oscilloscope and voltage source.

To measure ac voltage:
1. Set up calibrated oscilloscope.
2. Switch on internal sweep.

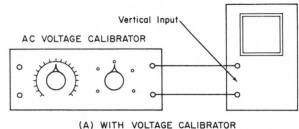

(A) WITH VOLTAGE CALIBRATOR

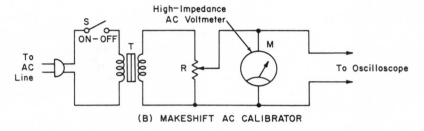

(B) MAKESHIFT AC CALIBRATOR

Fig. 4-4. Ac calibration with an external source.

3. Set SYNC switch to INTERNAL.
4. Set VERTICAL GAIN control to position used in calibration.
5. Connect VERTICAL INPUT terminals to test-voltage source.
6. Adjust sweep frequency for several cycles on screen.
7. Adjust HORIZONTAL GAIN control to spread pattern over as much of screen as desired. Pattern will resemble Fig. 4-1C.
8. Count number of screen divisions between positive peak (tip) and negative peak.
9. Determine voltage by multiplying number of divisions by calibration figure (peak-to-peak volts/division). This gives peak-to-peak value of unknown voltage.
10. If voltage is sinusoidal, multiply this value by 0.5 for peak voltage, by 0.3535 for rms voltage, or by 0.318 for average voltage.
11. If a low-capacitance probe is attached to oscilloscope, multiply indicated voltage by probe ratio. For example, if probe reduction ratio is 10:1, multiply indicated voltage by 10.

To measure dc voltage:

1. Set up calibrated dc oscilloscope.
2. Switch on internal sweep.
3. Set SYNC switch to INTERNAL.
4. Set VERTICAL GAIN control to position used during calibration.
5. Adjust HORIZONTAL GAIN control to lengthen horizontal-line trace on screen.

6. Connect VERTICAL INPUT terminals to test-voltage source.
7. Count number of divisions over which line was moved, up or down, from the zero line by test voltage.
8. Determine voltage by multiplying number of divisions by calibration figure (V/div.). Positive voltage deflects trace upward; negative voltage deflects it downward.

4.3 Voltage Measurement with Voltage Calibrator

The advantage of this method is that it does not require a precalibrated screen, and the vertical gain control may be set anywhere in its range. The voltage calibrator may be either internal or external.

Follow this procedure:
1. Set up oscilloscope.
2. Set SYNC switch to INTERNAL.
3. Apply signal to oscilloscope vertical input.
4. Set internal sweep frequency for desired number of signal cycles in pattern.
5. Set SYNC control for stationary pattern.
6. Adjust VERTICAL GAIN control for desired height and HORIZONTAL GAIN control for desired width of pattern.
7. Carefully note peak-to-peak height of pattern.
8. Without disturbing vertical gain, switch on calibrator. Signal disappears and is replaced by square-wave calibration pattern on screen.
9. Adjust CALIBRATOR control(s) until square wave fills same vertical space that signal filled.
10. Read peak-to-peak voltage from setting of calibrator control(s).

Figure 4-5 shows how the signal and calibrator voltages are compared on the screen in this manner. In Fig. 4-5A, the signal is a damped wave; its maximum peak-to-peak amplitude is to be measured. The maximum

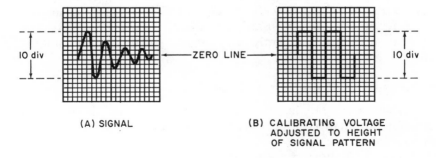

Fig. 4-5. A comparison of signal and calibrating voltages.

76 Practical Oscilloscope Handbook

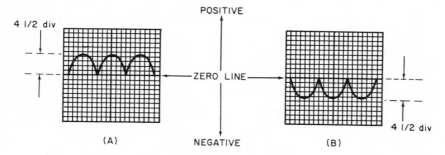

Fig. 4-6. Pulsating voltages.

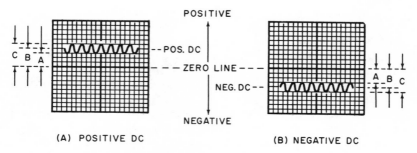

Fig. 4-7. Fluctuating voltages.

deflection is seen to occupy 10 divisions. After the calibrator is switched in, the square-wave voltage is adjusted for the same deflection (see Fig. 4-5B). The unknown voltage is then read from the settings of the calibrator control(s).

When you are working with dc signals, both the signal and calibrator voltages will be represented by a deflected dot (Fig. 4-3C or E) or deflected line (Fig. 4-3D or F).

With this method, a signal display may be interrupted as often as necessary and its voltage quickly checked, after which the signal display may be restored to the screen and the calibrator voltage removed. The technician does this frequently while making observations with the oscilloscope.

4.4 Measuring Pulsating Voltage

Pulsating dc contains only positive half-cycles or negative half-cycles. The output of a rectifier has such a waveform. So does the output of a dc generator.

Figure 4-6A shows a positive pulsating voltage; Fig. 4-6B shows a negative pulsating voltage. Note in each instance that the peak amplitude is four and one-half divisions. This amplitude may be evaluated by determining what voltage corresponds to four and one-half divisions.

Direct Method. Use a calibrated screen according to the method explained in Section 4.2. The value obtained will be *peak* voltage (since only one peak is present: positive in Fig. 4-6A, negative in Fig. 4-6B), although the screen calibration is given in peak-to-peak voltage.

Indirect Method. Use a voltage calibrator according to the method previously explained. If calibrator is ac type, the voltage value obtained will be peak-to-peak and must be divided by two to give the peak value of the pulsation. If calibrator is dc type, the voltage value obtained (read from dc scale of calibrator) equals peak value of the pulsation, and no calculation is required.

Increased stability of the zero line will be secured if a dc oscilloscope is used for these measurements.

4.5 Measuring Fluctuating (Composite) Voltage

A fluctuating voltage, also called composite voltage, consists of an ac superimposed upon a dc. Plate voltage of a tube amplifying an ac signal, or collector voltage of a transistor amplifying an ac signal is fluctuating. This type of voltage is measured best with a dc oscilloscope.

Figure 4-7 shows displays obtained with fluctuating voltages. In Fig. 4-7A, the positive dc component of the fluctuating voltage produces a deflection to the fourth line above zero. The ac component swings this voltage above the dc line and below the dc line (one division in each direction); i.e., from the 5th line (positive half-cycle) to the 3rd line (negative half-cycle). In Fig. 4-7B, the ac and dc voltage components are the same as in the preceding example, but the dc component is negative and this deflects the entire pattern downward by the same amount as before.

Measure the various components of the fluctuating voltage by either (1) the direct method, using a dc-calibrated screen (Section 4.2) or (2) the indirect method, using a dc voltage calibrator (Section 4.3) to check points A, B, and C. In Fig. 4-7A, the dc voltage corresponds to four divisions, and the peak-to-peak ac voltage to two divisions (i.e., from a deflection of five div. to the tip of the positive half-cycle to three div. to the tip of the negative half-cycle). If the direct screen calibration were 10 V/div., for example, the following values would be obtained: dc component +40 V, ac component 20 V p-p (7.07 V rms). Values are numerically the same in Fig. 4-7B, but the polarity is reversed; i.e., the dc component is –40 V, and the ac component oscillates about negative values.

4.6 Power Supply Ripple

Ripple in the dc output of an ac-operated power supply (including the vibrator type and electronic inverter type) is due to the ac supply. Its

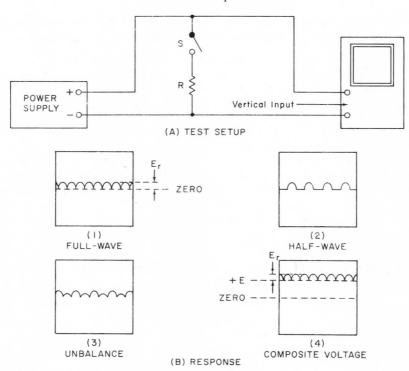

Fig. 4-8. Power-supply ripple.

amplitude and frequency depend upon whether the rectifier circuit is half-wave or full-wave and whether the ac supply is single-phase or polyphase. Measurement is similar to that of commutator ripple (described in Section 4.1).

Figure 4-8A shows the test setup. An ac oscilloscope is used to block the mean dc output voltage of the power supply from the oscilloscope (this voltage is high, compared to the ripple, and can cause off-screen deflection), and the internal recurrent sweep is operated. The ripple-voltage pattern resembles Fig. 4-8B1 for full-wave rectification, and Fig. 4-8B2 for half-wave rectification. These wave shapes apply both to single-phase and polyphase operation. The test is made under conditions of no load (switch S open) and full load (switch S closed).

Test Procedure

1. Set up voltage-calibrated ac oscilloscope.
2. Switch on internal sweep.
3. Set **SYNC SELECTOR** switch to **INTERNAL**.
4. Set **HORIZONTAL** and **VERTICAL GAIN** controls to midrange.

5. Connect equipment as shown in Fig. 4-8A. Select resistance R for full load of power supply. Power supply output (dc voltage plus peak ripple voltage) must not exceed maximum safe input voltage rating of oscilloscope.
6. Switch on power supply, noting that pattern appears on screen.
7. Adjust SWEEP FREQUENCY and SYNC controls for several, stationary ripple humps (Fig. 4-8B1 for full-wave supply; Fig. 4-8B2 for half-wave supply).
8. Readjust HORIZONTAL and VERTICAL GAIN controls, if necessary, for suitable width and height of pattern.
9. From voltage-calibrated vertical axis, measure peak amplitude of ripple voltage (see E_r in Fig. 4-8B1). Make separate measurements with switch S open and closed.
10. Determine ripple frequency by (a) adjusting SWEEP FREQUENCY and SYNC controls for a single, stationary ripple hump on screen, and (b) reading frequency from calibrated sweep. If oscilloscope does not have frequency-calibrated sweep, use an external audio oscillator (connected to horizontal input) and Lissajous figures to measure frequency.

If the halves of a full-wave circuit are unbalanced (mismatched rectifiers, one defective rectifier, unequal loading, breakdown on one side, etc.), the humps will alternate in height, as shown in Fig. 4-8B3. If the half-wave pattern (Fig. 4-8B1) is obtained from a full-wave power supply, one-half of the circuit is dead. Electrical noise generated by the power supply will produce fluctuations and noise waves on the ripple trace.

Sometimes, it is desired to observe output voltage and ripple simultaneously. This is a composite voltage measurement and requires a dc oscilloscope. Circuit connections are the same as in Fig. 4-8A. The dc output of the power supply deflects the base line upward from zero to +E (equal to the output voltage), as shown in Fig. 4-8B4. Ripple voltage E_r is then measured from the +E line to the peaks of the humps. However, low ripple is difficult to measure when +E is very high compared to E_r.

4.7 Measuring Ac and Dc Current

Current may be checked with a voltmeter by measuring the voltage drop (E) produced by the unknown current (I) flowing through an accurately known resistance (R), and calculating $I = E/R$. The oscilloscope is used in the same way: to measure the voltage drop across a shunt resistor.

Figure 4-9 shows the equipment setup for current measurement. To simplify calculations, a 1-ohm resistor is used. This must be a noninductive unit with a wattage rating numerically equal to twice the square of the maximum current (in amperes) to be measured.

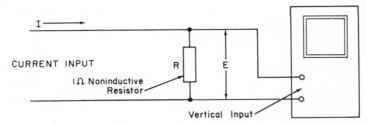

Fig. 4-9. The correct setup for current measurement.

To measure current:

1. Set up oscilloscope.
2. Connect 1-ohm shunt resistor (R), as shown in Fig. 4-9.
3. Pass unknown ac or dc current through R.
4. Measure voltage drop (E) across R, using either the direct or calibrator method explained. For peak current determine peak voltage; for rms current, determine rms voltage; for average current, determine average voltage.
5. From this, calculate the current: I = E, where I is in amperes and E is in volts.

If the current is pulsating, use the pulsating voltage measurement technique.

4.8 Measuring Fluctuating (Composite) Current

If the current is fluctuating:

1. Use the voltage measurement method, checking first the dc voltage drop (E_{dc}) and then the ac voltage drop (E_{ac}).
2. Convert the ac voltage to the rms value (E_{ac}).
3. Calcuate the ac and dc currents separately:

$$I_{ac} = E_{ac}, \text{ and } I_{dc} = E_{dc}$$

4. Finally, calculate the total current:

$$I_t = \sqrt{E_{ac}^2 + E_{dc}^2}$$

4.9 Current by Probe Method

Figure 4-10 is a simplified diagram of the current probe. This device resembles the familiar clamp-type ammeter used in power-frequency electrical measurements and operates in a somewhat similar manner, but the clamp transformer element is much smaller. For simplicity, the probe handle is not shown here, only the pickup element, which is mounted on its end.

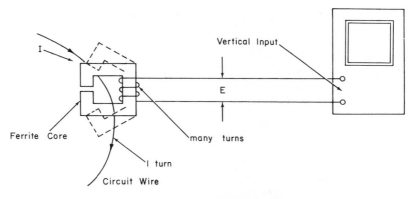

Fig. 4-10. Use of the current probe.

The pickup element consists of a tiny square-O-shaped ferrite core around which a coil of wire is wound. The core is swiveled so that it may be temporarily opened up, as shown by the dotted lines, when the probe handle is squeezed. This allows the core to be closed around a circuit wire carrying the current of interest. This forms a transformer with the wire acting as a one-turn primary against the probe coil which forms a many-turn secondary. The current, I, induces a voltage, E, across the secondary proportional to the step-up turns ratio, and this voltage is applied to the vertical amplifier input of a voltage-calibrated oscilloscope.

4.10 Comparing Two Waveforms on a Dual-Trace Scope

In observing simultaneous waveforms on channels A and B, it is necessary that the waveforms be related in frequency or that one of the waveforms be synchronized to the other although the basic frequencies may be different. An example of this is in checking a frequency divider or multiplier. The reference, or "clock," frequency can be used on Channel A, for example, and the multiple or submultiple of this reference frequency will be displayed on Channel B. In this way, when the waveform display of Channel A is synchronized, the display on Channel B will also be in sync with the Channel A display. If two waveforms having no phase or frequency relationship to each other are displayed simultaneously, it will be difficult if not impossible to lock both waveforms in sync for any useful observation.

Measurement Procedure

1. Perform the steps of the "Initial Starting Procedure," Section 2.26.
2. Connect oscilloscope probe cables to both the CH A and CH B INPUT (see Fig. 4-11).

82   *Practical Oscilloscope Handbook*

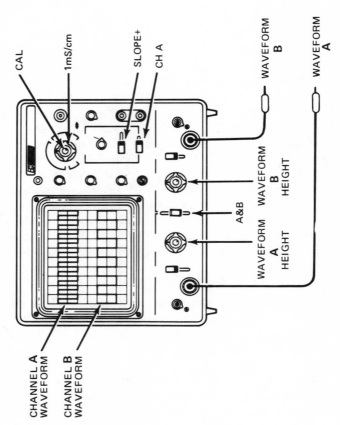

Fig. 4-11. Comparing two waveforms on a dual-trace scope. *(Courtesy Dynascan Corp.)*

Voltage and Current Measurement 83

3. Whenever possible, use the high-impedance, low-capacity 10:1 position to minimize circuit loading.
4. Set MODE switch to the A & B position. Two traces should appear on the screen.
5. Adjust CH A and CH B POSITION controls to place the Channel A trace above the Channel B trace, and adjust both traces to a convenient reference mark on the scale.
6. Set both the CH A and CH B DC-GND-AC switches to the AC position. This is the position used for most measurements and must be used if the points being measured include a large dc component.
7. Connect the ground clips of the probes to the chassis ground of the equipment under test. Connect the tips of the probes to points in the circuit where the waveforms are to be measured. It is preferred that the signal to which the waveform will be synchronized be applied to the Channel A input. If the equipment under test is a transformerless ac unit, use an isolation transformer to prevent dangerous electrical shock. The peak-to-peak voltage at the point of measurement should not exceed 600 V, if the probe is used in the DIRECT position.
8. Set the VOLTS/CM controls for Channels A and B to a position that gives 2 to 3 cm vertical deflection. The displays on the screen will probably be unsynchronized. The remaining steps, although similar to those outlined for single-trace operation, describe the procedure for obtaining stable, synchronized displays.
9. Set the SOURCE switch to the CH A position. This provides internal sync so that the Channel A waveform being observed is also used to trigger the sweep. If desired, the Channel B waveform may be used to trigger the sweep by setting the SOURCE switch to the CH B position. Often in dual-trace operation, a sync source other than the measurement point for Channel A or B is required. In this case, set the SOURCE switch to the EXT (external) position and connect a cable from the EXT TRIG jack to the sync source.
10. Set the SYNC switch to the TV(+) or TV(-) positions for observing television composite video waveforms, or to the SLOPE(+) or SLOPE(-) positions for observing all other types of waveforms. Use the (+) positions if the sweep is to be triggered by a positive-going wave, or the (-) positions if the sweep is to be triggered by a negative-going wave.
11. Adjust TRIGGERING LEVEL control to obtain a stable, synchronized sweep. As a starting point, the control may be pushed in and rotated to any point that will produce a sweep, which is usually somewhere in the center portion of its range. The trace will disappear if there is inadequate signal to trigger the sweep, such as when measuring extremely low-amplitude signals. If no sweep can be obtained, switch to automatic triggering.

12. Set SWEEP TIME/CM switch and VARIABLE control for the desired number of waveforms. These controls may be set for viewing only a portion of a waveform, but the trace becomes progressively dimmer as a smaller portion is displayed.
13. After obtaining the desired number of waveforms as in step 12, it is sometimes desirable to make a final adjustment of the TRIGGERING LEVEL control. The (-) direction of rotation selects the most negative point on the sync waveform at which sweep triggering will occur, and the (+) direction selects the most positive point on the sync waveform at which sweep triggering will occur. The control may be adjusted to start the sweep on any desired portion of the sync waveform.
14. The observed waveforms of channels A and B can be expanded by a factor of 5 by pulling outward on the ◄ ► POSITION control. This control can then be rotated clockwise or counterclockwise to view the left and right extremes of the waveform displays as desired.
15. Calibrated voltage measurements, calibrated time measurements, and operation with Z-axis input are identical to those described for single-trace operation in Section 2.26. Either the Channel A or Channel B vertical adjustment controls can be used as required in conjunction with the horizontal sweep controls to obtain the required amplitude or time interval measurements. This can be done either by using the dual display facilities such as the A & B position of the MODE switch or by reverting to single-trace operation, using the CH A or CH B positions of the MODE switch.
16. The Channel A and Channel B waveform displays can be added algebraically by placing the MODE switch in the A+B position, or subtracted algebraically in the A-B position.

4.11 Differential Voltage Measurement Using Dual-Trace Scope

A dual-trace oscilloscope is convenient to observe waveforms and measure voltages between two points in a circuit, neither of which is circuit ground (Fig. 4-12). Such measurements as the inputs to a differential amplifier, the output of a phase splitter or push-pull amplifier, the amount of signal developed across a single section of voltage divider or attenuator, and many others require this technique.

Measurement Procedure

1. Adjust controls as previously described under "Initial Starting Procedure," in Sections 2.26.
2. Connect a probe cable to both the CH A and CH B INPUT jacks (Fig. 4-12).
3. Connect ground clips of the two probes to the chassis of equipment under test, and connect tips of the probes to the points in the circuit

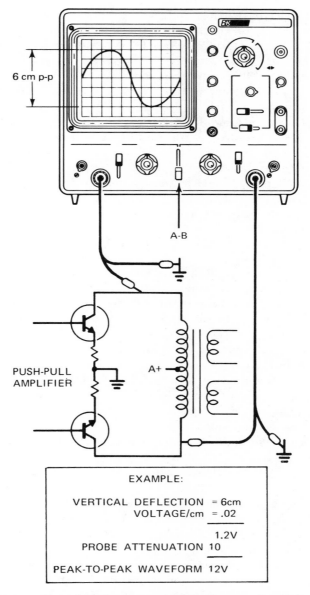

Fig. 4-12. Differential voltage measurement. *(Courtesy* Dynascan Corp.)

where measurements are to be made. It is usually desirable to connect the CH A probe to the higher potential or higher amplitude point in the circuit and the CH B probe to the lower potential or lower amplitude point in the circuit.

4. Set the MODE switch to the CH A position and the SOURCE switch to the CH A position and adjust the controls as previously instructed in the "Single-Trace Waveform Observation" procedure to obtain a synchronized single waveform of 2 to 6 cm vertical height with the CH A VARIABLE control set to CAL.
5. If only the AC component of the waveform is of interest, use the following procedure:
    a. Set CH A and CH B DC-GND-AC switch both to the AC position.
    b. Set CH A VARIABLE control to CAL and the CH B VARIABLE control to CAL and the CH B VOLTS/CM switch to the same position as the CH A VOLTS/CM switch.
    c. If the Channel A and Channel B inputs are in phase, set the MODE switch to the A−B position. The displayed waveform is the peak-to-peak difference between the two points of measurement. If the waveform is small, the vertical sensitivity may be increased but the CH A and CH B VOLTS/CM switches must both be in the same position.
    d. If the Channel A and Channel B inputs are 180° out of phase, such as the output of a push-pull amplifier, set the MODE switch to the A−B position to measure the full peak-to-peak waveform. Set the MODE switch to the A+B position to measure any imbalance between the two points of measurement. Readjust the VOLTS/CM switches as required to obtain as large a waveform as possible without exceeding the limits of the vertical scale, but always keep the CH A and CH B switches set to the same sensitivity.
    e. Position the waveform as desired with the positioning controls and calculate the peak-to-peak voltage as described in the "Calibrated Voltage Measurement" procedure, Section 2.28.
6. If a dc voltage, or the dc component of the waveform is of interest, use the following procedure:
    a. Set CH A DC-GND-AC switch to the DC position.
    b. Position the CH A VOLTS/CM switch to keep the trace within the limits of the vertical scale. Use the CH A POSITION control to align the trace with one of the lines on the scale for reference.
    c. Position CH B VOLTS/CM switch to the same position as the CH A VOLTS/CM switch.
    d. Set CH B DC-GND-AC switch to the GND position and adjust out any error that may be introduced by the Channel B positioning control as follows: Alternately set the MODE switch to the A+B and A−B positions adjusting the CH B POSITION control until the trace position does not shift as the MODE switch position changes.
    e. Return CH B DC-GND-AC switch to the DC position.

f. Momentarily return the MODE switch to the CH A position and note the trace position for reference. You may readjust it with the Channel A vertical positioning control, but not the Channel B control. Place the MODE switch in the A-B position. The amount of displacement of the trace from the Channel A reference represents the voltage differential between the two points of measurement.

5

frequency and phase measurement and comparison

The oscilloscope is a sensitive indicator in frequency and phase checking because of its utilization of the patterns known as *Lissajous figures*. (It is used for frequency measurement by other methods also.) The techniques are simple and the results are dependable. Measurements may be made at any frequency in the response range of the oscilloscope. High-amplitude signals outside of the amplifier passband may be applied directly to the deflecting plates.

5.1 Use of Lissajous Figures

The use of Lissajous figures permits the comparison of one frequency with another. The unknown frequency may be measured in terms of a known one, or one frequency may be adjusted to equal the other with no knowledge of the values of either one.

One frequency is applied to the oscilloscope horizontal channel, the other to the vertical channel, as shown in Fig. 5-1. Either signal may be applied to either channel; commonly, however, the unknown is presented to the vertical, and the known (standard) to the horizontal. For convenience in the following explanations, the horizontal signal is designated f_h, and the vertical signal is designated f_v.

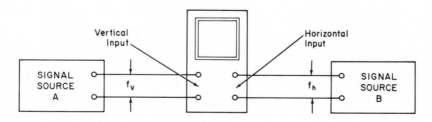

Fig. 5-1. The test setup for Lissajous figures.

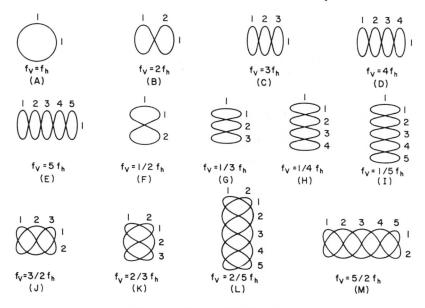

Fig. 5-2. Typical Lissajous figures.

Measurement Procedure

1. Set up oscilloscope.
2. Switch off internal sweep.
3. Switch off sync.
4. Connect signal sources to oscilloscope, as shown in Fig. 5-1.
5. Advance HORIZONTAL and VERTICAL GAIN controls, noting that pattern (very likely spinning) appears on screen.
6. Set HORIZONTAL and VERTICAL GAIN controls for desired width and height of pattern.
7. While holding frequency f_v constant, vary frequency f_h, noting that pattern spins in alternate directions and changes shape.
8. The pattern stands still whenever f_h and f_v are in integral ratio (either even or odd).
9. When $f_h = f_v$, pattern stands still and (if each signal is sinusoidal) is a single circle or ellipse (Fig. 5-2A).
10. When $f_v = 2f_h$, a two-loop horizontal pattern (Fig. 5-2B) appears.

To determine the frequency from any Lissajous figure, count the number of horizontal loops in the pattern and use this number for the numerator of a fraction; then count the number of vertical loops and use that number as the denominator; and finally multiply f_h (known frequency) by the fraction. Thus, in Fig. 5-2H, there are 1 horizontal loop and

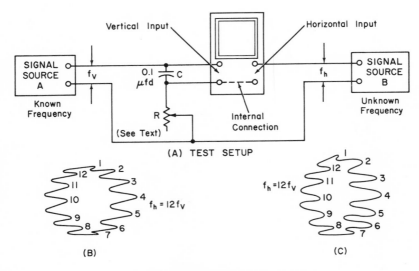

Fig. 5-3. The modulated-ring method.

4 vertical loops, giving a fraction of ¼. The unknown frequency (f_v), therefore, is ¼f_h. Figure 5-2 shows some common Lissajous figures, with the loops numbered for illustrative purposes.

An accurately calibrated, variable-frequency oscillator will supply the horizontal search frequency for frequency measurement. When two frequencies are to be matched, and absolute frequency value is of no interest, either one may be applied to the vertical channel, and the variable-frequency one adjusted for the 1:1 pattern in Fig. 5-2A.

Use of Lissajous figures is limited by the number of loops which can be observed and counted on the screen. As the number increases, counting becomes progressively difficult. Accuracy is enhanced by a large screen, sharp focusing, steady display, and keen eyesight.

5.2 Use of Modulated-Ring Pattern

When a Lissajous figure contains a large number of loops, accurate counting becomes difficult. Figure 5-3 shows a test method that uses a modulated-ring pattern in place of the looped figure and permits a higher count. This pattern (Fig. 5-3B and C) is also called a *gear wheel* or *toothed wheel*, from its shape. The unknown frequency is determined by multiplying the known frequency by the number of teeth in the pattern. A large number of teeth may be formed on a circle spread over most of the screen.

Figure 5-3A shows the equipment setup. Here, a phase-shift network (RC) introduces a 90° phase shift between the horizontal and vertical channels of the oscilloscope that is needed to produce a circle (ring) pattern with the known frequency f_v. Voltage from the unknown frequency source modulates this ring, as shown in Fig. 5-3B. When the voltages across the capacitor and resistor are not equal, the pattern is elliptical, instead of circular, as shown in Fig. 5-3C. The unknown frequency must be higher than the known frequency, and the amplitude of the unknown must be reduced below that of the known to prevent distortion of the pattern. For the required voltages across R and C to be equal the resistance of R must equal the reactance of the 0.1-μF capacitor C at the frequency of signal source A. Making A variable enables exact setting for a smooth circle.

Measurement Procedure

1. Set up oscilloscope.
2. Switch off internal sweep.
3. Switch off sync.
4. Connect equipment as shown in Fig. 5-3A, but temporarily switch off unknown signal source B.
5. Switch on signal source A.
6. Adjust R for a ring pattern on screen.
7. Adjust HORIZONTAL and VERTICAL GAIN controls to spread ring over maximum usable area of screen.
8. Switch on signal source B, noting that ring becomes wrinkled or toothed by unknown signal.
9. Adjust known frequency f_v to stop ring from spinning.
10. Adjust amplitude of f_h voltage for distinct teeth on pattern.
11. Count number (n) of teeth on pattern.
12. Calculate unknown frequency:

$$f_h = nf_v$$

Unless f_h is an integral multiple (even or odd) of f_v, the wheel will spin counterclockwise or clockwise when f_h is not an exact multiple of f_v. Adjust the known frequency to stop the wheel.

5.3 Use of Broken-Ring Pattern

When the oscilloscope has a Z-axis input, a circular pattern may be obtained that is broken into segments (rather than wrinkled) by the unknown signal cycles. Figure 5-4B shows such a pattern (sometimes called *spot wheel* or *dot wheel*). Either the segments or the holes (whichever is more distinct) are counted, and the known frequency is multiplied by this number to determine the unknown frequency. The number of recognizable segments obtained on a circle of given circumference is usually greater than the number of teeth obtained with the modulated-ring method. As in the latter method, the unknown frequency must be higher than the known frequency.

Figure 5-4A shows the equipment setup. The known signal (f_s) from signal source A is applied to the horizontal and vertical channels of the

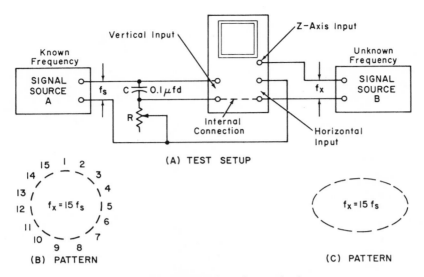

Fig. 5-4. The broken-ring method.

oscilloscope through a 90° phase-shift network, RC. This produces a smooth ring pattern on the screen. The unknown signal (f_x) from signal source B is applied to the Z-axis input, and, if of sufficient amplitude, punches a hole in the ring for each negative half-cycle of f_x. The positive half-cycles brighten the segments. The result is the broken-ring pattern shown in Fig. 5-4B.

For a circle pattern, resistance R must equal the reactance of 0.1 μF capacitor C at frequency f_s. Making R variable enables exact setting of the applied horizontal and vertical voltages for a smooth circle. If the reactances are not equal, the pattern will be elliptical, as in Fig. 5-4C.

Measurement procedure is similar to the modulated-ring method except that the INTENSITY control is adjusted for sharp contrast between segments and holes in the pattern. Instead of counting the number of teeth on the pattern, you count the number of segments or holes.

As with the modulated ring, unless f_x is an integral multiple of f_s, the wheel will spin counterclockwise or clockwise when f_x is not an exact multiple of f_s. Adjust the known frequency to stop the spin.

5.4 Use of Broken-Line Pattern

A variation of the previous method gives a straight-line trace broken into segments by the unknown frequency. This scheme is simpler, since it requires no phase-shift network, but it does not permit as high a count as the modulated-ring and broken-ring patterns.

In this method, the known frequency (f_s) is applied to the horizontal channel of the oscilloscope and produces a straight, horizontal line trace.

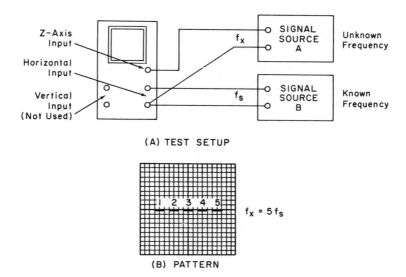

Fig. 5-5. The broken-line method.

The unknown frequency (f_x) is applied to the Z-axis input and, if of sufficient amplitude, punches a hole in the line for each negative half-cycle of f_x. The positive half-cycles brighten the segments. This results in the broken-line pattern shown in Fig. 5-5B.

Measurement Procedure

1. Set up oscilloscope.
2. Switch off internal sweep.
3. Switch off sync.
4. Set VERTICAL GAIN control to zero.
5. Connect equipment as shown in Fig. 5-5A, but temporarily switch off unknown signal source A.
6. Switch on known signal source B.
7. Adjust HORIZONTAL GAIN control to spread resulting horizontal straight-line trace over maximum usable width of screen.
8. Switch on signal source A, noting that line becomes broken into segments by unknown signal.
9. Adjust known frequency f_s to stop segments from drifting.
10. Adjust amplitude of f_x voltage for clean segmentation of line.
11. Adjust INTENSITY control for sharp contrast between segments and holes in pattern.
12. Count number (n) of segments.
13. Calculate unknown frequency:

$$f_x = nf_s$$

Unless f_x is an integral multiple of f_s, the segments will drift to the left or to the right. Adjust the known frequency to freeze the pattern.

5.5 Use of Sawtooth Internal Sweep

If the SWEEP FREQUENCY controls of the oscilloscope are direct reading in frequency or time, they may be used to identify an unknown frequency.

Connect the unknown signal to the VERTICAL INPUT. Set up the oscilloscope to obtain a single stationary cycle on the screen. Read unknown frequency from the scale of the SWEEP FREQUENCY or TIME CONTROL.

5.6 Calibrated Time Measurements

Pulse width, waveform periods, circuit delays, and all other waveform time durations are easily and accurately measured. Calibrated time measurements from 0.5 sec down to 0.1 μsec are possible. At low sweep speeds, the entire waveform is not visible at one time. However, the bright spot can be seen moving from left to right across the screen, which makes the beginning and ending points of the measurement easy to spot.

Measurement Procedure

1. Adjust controls as described in Section 2.3 for a stable display of the desired waveform.
2. Be sure the sweep time VARIABLE control is fully clockwise to the CAL position.
3. Set the SWEEP TIME/CM control for the largest possible display of the waveform segment to be measured, usually one cycle.
4. If necessary, readjust the TRIGGERING LEVEL control for the most stable display.
5. Read the amount of horizontal deflection (in centimeters) between the points of measurement.
6. Calculate the time duration as follows: Multiply the horizontal deflection (in centimeters) by the SWEEP TIME/CM switch setting. Remember, when the 5X magnification is used, the result must be divided by 5 to obtain the actual time duration. See example shown in Fig. 5-6.
7. Time measurements often require external sync. This is especially true when measuring delays. The sweep is started by a sync signal from one circuit and the waveform measured in a subsequent circuit. This allows measurement of the display between the sync pulse and the subsequent waveform. To perform such measurements using external sync, use the following steps:
   a. Set the SOURCE switch to the EXT position.

Frequency and Phase Measurement and Comparison

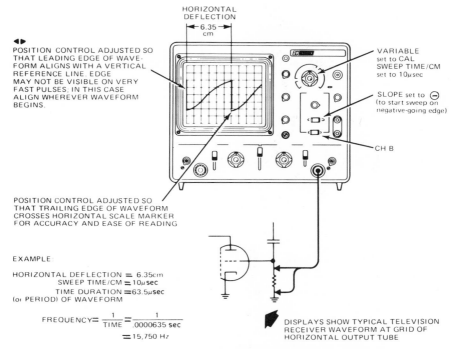

Fig. 5-6. Calibrated time measurements. *(Courtesy* Dynascan Corp.)

b. Connect a short length of shielded cable from the EXT TRIG jack to the source of sync signal.
c. Set the SYNC switch to the SLOPE (+) or (-) position for the proper polarity for the sync signal.
d. Readjust the TRIGGERING LEVEL control if necessary for a stable waveform.
e. If measuring a delay, measure the time from the start of the sweep to the start of the waveform.

5.7 Using a Dual-Trace Scope for Frequency Divider Analysis

Figure 5-7 illustrates the waveforms involved in a basic divide-by-two circuit. Figure A indicates the reference or "clock" pulse train. Figures B and C indicate the possible outputs of the divide-by-two circuitry. Figure 5-7 also indicates the settings of specific oscilloscope controls for viewing these waveforms. In addition to these basic control settings, the TRIGGERING LEVEL control as well as the Channel A and Channel B vertical position controls should be set as required to produce suitable displays. In the drawing of Fig. 5-7, the waveform levels of 2 cm are indicated. If the exact voltage amplitudes of the Channel A and Channel B are desired, the

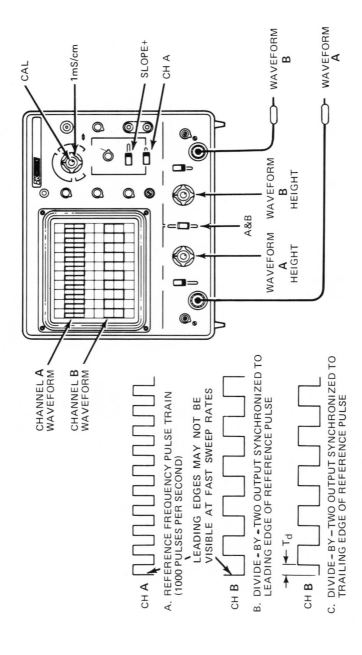

Fig. 5-7. Checking divide-by-two circuit. (*Courtesy Dynascan Corp.*)

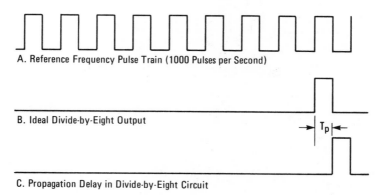

Fig. 5-8. Waveforms in a divide-by-eight circuit. *(Courtesy* Dynascan Corp.)

Channel A and channel B VARIABLE controls must be placed in the CAL position. The Channel B waveform may be that indicated either in Fig. 5-7B or in Fig. 5-7C. In Fig. 5-7C the divide-by-two output waveform is shown for the case where the output circuitry responds to a negative-going waveform. In this case, the output waveform is shifted with respect to the leading edge of the reference frequency pulse by a time interval corresponding to the pulse width.

5.8 Divide-by-Eight Circuit Waveforms with a Dual-Trace Scope

The waveform relationships for a basic divide-by-eight circuit are shown in Fig. 5-8. The basic oscilloscope settings are identical to those used in Fig. 5-7. The reference frequency of Fig. 5-8A is supplied to the Channel A input, and the divide-by-eight output is applied to the Channel B input. Figure 5-8B indicates the ideal time relationship between the input pulses and the output pulse.

In an application where the logic circuitry is operating at or near its maximum design frequency, the accumulated risetime effects of the consecutive stages produce a built-in time propagation delay that can be significant in a critical circuit and must be compensated for. Figure 5-8C indicates the possible time delay that may be introduced into a frequency divider circuit. By use of the dual-trace oscilloscope, the input and output waveforms can be superimposed to determine the exact amount of propagation delay that occurs.

5.9 Propagation Time Measurement Using a Dual-Trace Scope

An example of propagation delay in a divide-by-eight circuit was given in Section 5.8. Significant propagation delay may occur in any

98　　　　　　　　*Practical Oscilloscope Handbook*

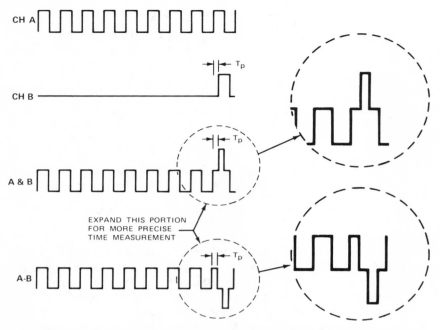

Fig. 5-9. Propagation-time measurements. *(Courtesy* Dynascan Corp.)

circuit with several consecutive stages. Figure 5-9 shows the resultant waveforms when the dual-trace presentation is combined into a single-trace presentation by selecting the A+B or A−B position of the MODE switch. In the A+B position, the two inputs are algebraically added in a single-trace display. Similarly, in the A−B position, the two inputs are algebraically subtracted. Either position provides a precise display of the propagation time (T_p). By using the procedures given for calibrated time measurement, (Section 5.6), T_p can be calculated. A more precise measurement can be obtained if the T_p portion of the waveform is expanded horizontally using the 5X multiplier.

5.10 Digital Circuit Time Relationships with a Dual-Trace Scope

A dual-trace oscilloscope is a valuable tool in designing, manufacturing, and servicing digital equipment; it permits easy comparison of time relationships between two waveforms.

In digital equipment, it is common for a large number of circuits to be synchronized, or to have a specific time relationship to each other. Many of the circuits are frequency dividers, and waveforms are often time-related in other combinations. In the dynamic state, some of the waveforms change, depending upon the input or mode of operation. Figure 5-10 shows a typical digital circuit and identifies several of the points at which waveform measurements are appropriate. The accompanying Fig. 5-11 shows the normal waveforms to be expected at the each of these points and their timing relationships. The individual waveforms have limited value unless their timing relationship to one or more of the other waveforms is

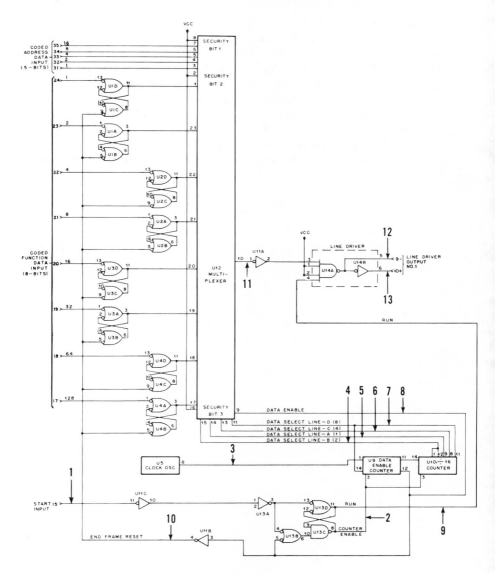

Fig. 5-10. Typical digital circuit with time-related waveforms. *(Courtesy Dynascan Corp.)*

known to be correct. In typical fashion, waveform No. 3 would be displayed on Channel A and waveforms No. 4 through 8 and No 10 would be successively displayed on Channel B, although other timing comparisons may be desired. Waveforms No. 11 through 13 would be displayed on Channel B in relationship to waveform No. 8 or 4 on Channel A.

In the family of time-related waveforms shown in Fig. 5-11, waveform No. 8 or 10 is an excellent sync source for viewing all of the waveforms; there is but one triggering pulse per frame. For convenience, external sync using waveform No. 8 or 10 as the sync source may be desirable. With external sync, any of the waveforms may be displayed without readjustment of the sync controls. Waveforms No. 4 through 7 should not be used as the sync source because they do not contain a triggering pulse at the start of the frame. It would not be necessary to view all the waveforms as shown in Fig. 5-11 in all cases. In fact, there are many times when a closer examination of a portion of the waveforms would be more appropriate. In such cases, it is recommended that the sync remain

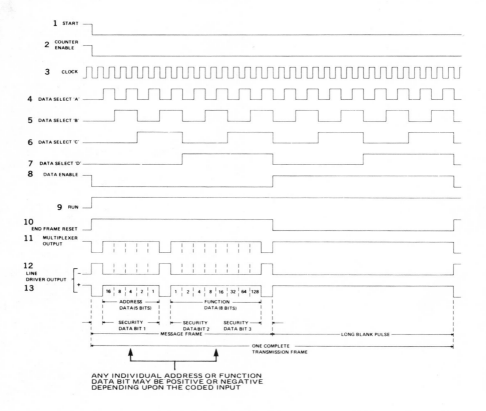

Fig. 5-11. Time-related waveforms for digital circuit shown in Fig. 5-10. (*Courtesy* Dynascan Corp.)

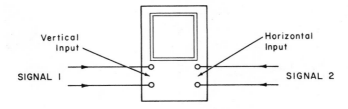

Fig. 5-12. The setup for phase measurement by Lissajous figures.

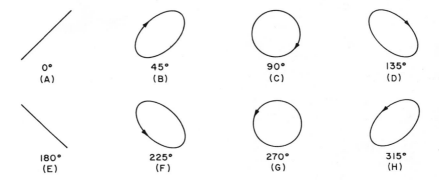

Fig. 5-13. Typical Lissajous figures for phase shift

unchanged while the sweep speed or 5X magnification be used to expand the waveform display.

5.11 Use of Lissajous Figures for Phase Measurement

When two signals are applied simultaneously to an oscilloscope without internal sweep, one to the horizontal channel and the other to the vertical channel, the resulting pattern is a *Lissajous figure* that shows phase difference between the two signals. Such patterns result from the sweeping of one signal by the other, and are similar to some of the Lissajous figures used for frequency measurement.

Figure 5-12 shows the test setup for phase measurements by means of Lissajous figures. Figure 5-13 shows patterns corresponding to certain phase difference angles when the two signal voltages are sinusoidal, equal in amplitude, and equal in frequency. Note that the same patterns sometimes are obtained for widely different angles: a right-tilted ellipse for both 45° and 315°. The spot, however is moving in a different direction for each: clockwise in Fig. 5-13B and counterclockwise in Fig. 5-13H. The angles shown are for the phase angle of the vertical input with respect to the horizontal input. A simple way to find the correct phase angle (whether leading or lagging) is to introduce a small, known phase shift to one of the inputs. The proper angle may then be deduced by noting the direction in which the pattern changes.

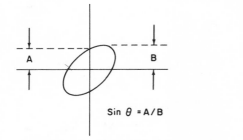 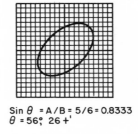

(A) MEASUREMENTS ON PATTERN

(B) PATTERN OF TYPICAL DIMENSIONS

Fig. 5-14. Determination of phase from Lissajous figures.

Measurement Procedure

1. Set up oscilloscope.
2. Switch off internal sweep.
3. Switch off sync.
4. Apply SIGNAL 1 and SIGNAL 2, as shown in Fig. 5-12.
5. Advance HORIZONTAL and VERTICAL GAIN controls for sample pattern on screen.
6. The two signals should be equal in amplitude. If they are not, either (a) adjust output controls in signal sources for equal signal voltages (with HORIZONTAL and VERTICAL GAIN controls set for identical gain); or (b) if signal sources have no output control, adjust HORIZONTAL and VERTICAL GAIN controls so that equal signal voltages are applied to the deflecting plates.
7. Using the scheme shown in Fig. 5-14, carefully measure vertical deflection from zero line to point at which pattern intersects center vertical line of screen. Record this dimension (in inches, centimeters, or scale divisions) as A.
8. Carefully measure maximum height of pattern from zero line, and record this dimension as B.
9. Calculate sine of phase difference angle:

$$\sin \theta = A/B$$

10. Find the corresponding angle in a Table of Sines.

It is seen from Fig. 5-14A that this method may be used to find the phase angle from any Lissajous figure obtained by the test method just described. Thus, the ellipse in Fig. 5-14B intersects the vertical axis at 5 divisions and has a maximum height of 6 divisions. From these screen measurements:

$$\sin \theta = 5/6 = 0.8333$$

A sine table will show that this corresponds to an angle of 56°, 26+'. The straight-line patterns (Figs. 5-13A and 5-13E) intersect the vertical axis at

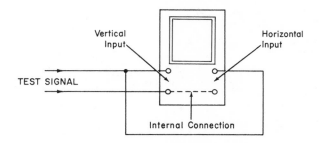

Fig. 5-15. The setup for checking the internal phase shift of an oscilloscope.

zero, hence A = 0. Their sine, consequently, is equal to 0/B = 0 (corresponding to 0° and 180°, respectively).

5.12 Checking Inherent Phase Shift of Oscilloscope

No oscilloscope should be used for phase shift measurements unless its own internal phase shift (the phase difference between horizontal and vertical amplifiers) has been checked first. This difference is held to a small figure in a good oscilloscope having identical horizontal and vertical channels.

Figure 5-15 shows the setup for checking internal phase shift. A test signal of desired frequency is applied simultaneously to horizontal and vertical amplifiers. The horizontal and vertical input signals are then in phase with each other, since they are the *same* signal. The resulting Lissajous figure is measured and the phase angle calculated.

Measurement Procedure

1. Set up oscilloscope.
2. Switch off internal sweep.
3. Switch off sync.
4. Apply test signal of desired frequency, as shown in Fig. 5-15.
5. Set HORIZONTAL and VERTICAL GAIN controls for equal amplification.
6. From Lissajous figure on screen, determine phase angle as explained in Section 5.11.
7. Repeat at as many frequencies as practicable throughout passband of oscilloscope. (Some instruments will show excessive phase shift at some frequencies but will be free of this trouble at others.)

5.13 Use of Dual Pattern

A dual-trace display offers the advantage that two signal waves and their phase relations may be observed directly without Lissajous figures. Neither the signal amplitudes nor frequencies need be equal, nor need either signal be sinusoidal.

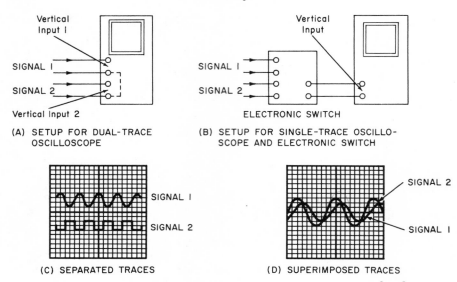

Fig. 5-16. The use of a dual pattern.

Figure 5-16 illustrates this method. In Fig. 5-16A, an oscilloscope is shown which has two-channel input. In Fig. 5-16B, a conventional oscilloscope is shown with an electronic switch, which enables two signals to be displayed simultaneously on the screen on a single-gun CRT. The signal-position control may be adjusted for separated traces (Fig. 5-16C) or superimposed traces (Fig. 5-16D).

In most cases, the phase relationships may be determined by inspection. Thus, in Fig. 5-16C, Signal 1 (sine wave) is seen to be 180° out of phase with Signal 2 (square wave), since one reaches its positive peak at the same instant that the other reaches its negative peak. In Fig. 5-16D, Signal 2 is seen to lead Signal 1 by one-eighth of a cycle (45°).

Measurement Procedure

1. Set up oscilloscope.
2. Switch on internal sweep.
3. Set SYNC SELECTOR switch to INTERNAL.
4. Apply test signals to oscilloscope, as shown in Figs. 5-16A or 5-16B.
5. Adjust SWEEP FREQUENCY and SYNC controls for several stationary cycles of each signal.
6. Adjust HORIZONTAL and VERTICAL GAIN controls for desired width and height of patterns.
7. Adjust SIGNAL POSITION control (in electronic switch or dual trace oscilloscope) to separate patterns (Fig. 5-16C) or superimpose them (Fig. 5-16D), as desired.

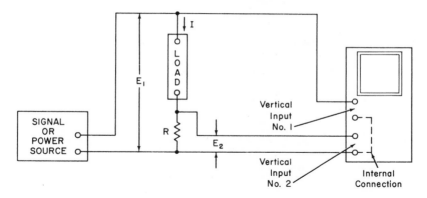

Fig. 5-17. The setup for checking the angle between current and voltage.

8. Determine phase difference by noting displacement of one signal with respect to the other along horizontal axis.

5.14 Checking Phase Angle between Current and Voltage

The preceding sections explained the measurement of phase angle between two voltages. In practice, it is often necessary to measure the angle between voltage and current. In this case, the current may be handled by converting it into a proportional voltage. This is done by passing it through a low, noninductive resistance and using the voltage drop across this resistance as the oscilloscope signal.

Figure 5-17 shows a typical setup. Either a dual-trace oscilloscope or single-trace oscilloscope with electronic switch must be used. Here, it is desired to know the phase relationship between voltage E_1 across the load and current I through the load. Resistance R is negligible with respect to the load impedance (generally it is between 1 and 10 ohms). Load voltage E_1 is applied to VERTICAL INPUT No. 1 to produce one pattern on the screen. Load current I sets up a voltage drop, E_2, across resistor R, and this voltage is applied to VERTICAL INPUT No. 2 to produce a second pattern, which is proportional to the current. From these two patterns, phase relations may be observed. Use the same procedure as previously described for dual pattern.

5.15 Checking Phase Angle between Two Currents

When the phase angle between two currents is to be checked, the currents are passed through separate low resistances, and the resulting voltage drops are applied to the two vertical inputs of a dual-trace oscilloscope or to the two signal inputs of an electronic switch operated into a conventional oscilloscope. The angle is then determined from the relationship of the two patterns.

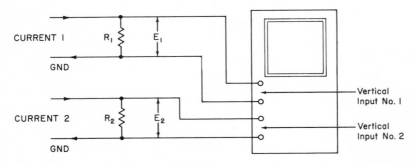

Fig. 5-18. The setup for checking the angle between two currents (common ground).

Figure 5-18 shows the test setup. Current 1 flows through resistor R_1 and sets up voltage drop E_1, which produces one pattern on the screen. Current 2 flows through resistor R_2 and sets up voltage drop E_2, which produces a second pattern. The height of each pattern is proportional to the corresponding current. The resistances are so low (1 ohm, for example) that their presence will not upset operation of the circuit to which the instrument is connected. Use the measurement procedure described for dual pattern.

6

audio amplifier, receiver, and transmitter tests and measurements

Engineer, laboratory aide, service technician, and audiophile alike find the oscilloscope a serviceable audio tool. Use of this single instrument can be informative, as well as a time and labor saver in design checking, troubleshooting, and maintenance.

6.1 Checking Wave Shape

The oscilloscope can quickly verify the waveform of an AF signal (whether it is sinusoidal or rectangular, sawtooth or steep pulse, pure or distorted, etc.). The instrument is often used to obtain a qualitative indication of this sort when type of wave—not frequency, voltage, or phase—is the chief matter of interest. It often is used directly to inspect the signal present in an amplifier or line, or delivered by a generator or a playback head. It is also used in conjunction with some other instrument, such as a voltmeter, current meter, distortion meter, or wave analyzer, to monitor waveform while the other instrument indicates a quantity.

Test Procedure

1. Set up oscilloscope.
2. Switch on internal sweep.
3. Set **SYNC SELECTOR** switch to **INTERNAL**.
4. Connect **VERTICAL INPUT** terminals to signal source. Use low-capacitance probe if minimum disturbance to source must be assured.
5. Adjust **SWEEP FREQUENCY** and **SYNC** controls for several stationary cycles on screen.
6. Adjust **HORIZONTAL** and **VERTICAL GAIN** controls for desired width and height of pattern.
7. Observe shape of signal pattern.

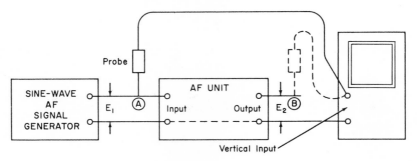

Fig. 6-1. The setup for checking voltage gain.

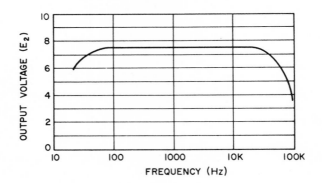

Fig. 6-2. A typical frequency response curve.

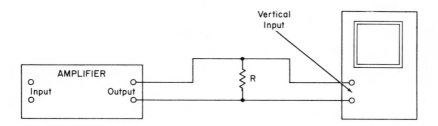

Fig. 6-3. The setup for the hum and noise test.

6.2 Checking Voltage Gain or Loss

In this application, the oscilloscope is used as a high-impedance electronic AF voltmeter. Its sensitivity in the millivolts region depends upon the gain (sensitivity) of the vertical amplifier. Figure 6-1 shows the test setup. Here, the *AF unit* may be a complete amplifier, one or more stages of an amplifier, a component (such as a transformer), or a network (such as a filter). The signal generator is a low-distortion source set to the desired test frequency.

If the unit under test is a power amplifier, connect a load resistor across the output terminals. Adjust the scope for several stationary cycles as described in Section 6.1. If signal peaks appear flattened (overloaded), reduce the signal generator level. First measure the voltage at point A, using one of the voltage-measuring techniques described in Chapter 4. Then place probe at point B and measure voltage.

Voltage gain $A = E_2/E_1$ and expressed in dB, $A = 20 \log_{10} E_2/E_1$.

6.3 Checking Frequency Response

This application demands that the oscilloscope response be excellent up to at least twice the highest test frequency to be used. The test method is simple. A constant-amplitude, sinusoidal, test-signal voltage is applied to the amplifier or component under test. The signal frequency is varied throughout the AF spectrum (e.g., 20 to 25,000 Hz) and the corresponding amplifier output voltages are checked at as many frequencies as practicable. The oscilloscope is used to check the input and output voltages, and a curve may be drawn to show variation of output voltage with frequency (Fig. 6-2).

Adjust the gain and tone controls of the amplifier to desired operating position. If the unit is a power amplifier, place a load resistor (equal to the output impedance and of sufficient wattage) across the output terminals. Use the same procedure as described in Section 8.2, but start at the lowest test frequency, complete the gain measurement, and then select a number of different frequencies up to and including the highest frequency to be checked.

6.4 Checking Hum and Noise Level

Measurement of residual hum and noise in an amplifier, under zero-signal conditions, requires a sensitive oscilloscope (Fig. 6-3). This is because of the low amplitude of hum and noise voltages (often less than 1 mV rms). The proper procedure is to check the output of the amplifier, with power switched on but with no input signal. Any output voltage under these conditions will be due to hum, noise, or self-oscillation. Use of the oscilloscope has the advantage that waveform, as well as voltage amplitude, may be observed, and the components may be identified by means of frequency measurements.

If a power amplifier is under test, use a load resistor of proper value and wattage across the output terminals. First, set the gain control to zero. The procedure described in Section 6.1 is used except that no input is applied. The output observed is thus due to hum and noise. The hum component can be viewed by setting the SWEEP FREQUENCY control to 60 Hz and the sync control for one stationary pattern. If oscillation is present, its amplitude and frequency can be measured using procedures previously described. Repeat with the gain control set to maximum.

6.5 Measuring Power Output

This is another application in which the oscilloscope is used essentially as an electronic AF voltmeter. Figure 6-4 shows the test setup.

Measurement Procedure

1. Set up oscilloscope.
2. Switch on internal sweep.
3. Set SYNC SELECTOR switch to INTERNAL.
4. Set up equipment as shown in Fig. 6-4. Resistance R must equal the normal load impedance of amplifier.
5. Switch on amplifier and generator, and set amplifier gain control to maximum-gain position. Set generator to desired test frequency.

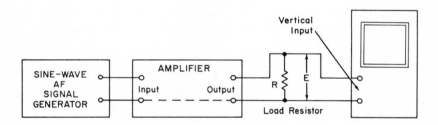

Fig. 6-4. The setup for power output measurement.

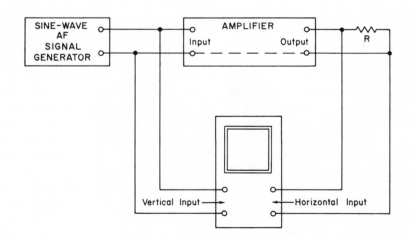

Fig. 6-5. The setup for phase-shift measurement.

6. Advance output of generator.
7. Adjust VERTICAL GAIN control until pattern appears on screen.
8. Adjust SWEEP FREQUENCY and SYNC controls for several stationary cycles.
9. If amplifier is overdriven, as evidenced by flattening of signal peaks, reduce generator output to remove this distortion.
10. Measure rms signal voltage (E_1), using one of methods outlined in Chapter 4.
11. Calculate power output:

$$P = E^2/R$$

where P is in watts, E in volts, and R in ohms.

12. Repeat power measurement at each desired test frequency within passband of amplifier.

6.6 Checking Amplifier Phase Shift

The phase shift introduced by a complete amplifier or by a single amplifier stage (or combination of stages) may be measured with the setup shown in Fig. 6-5.

The measurement procedure is similar to that described in Section 5.11.

Phase shift may also be checked with a dual-trace oscilloscope (or single-trace oscilloscope with electronic switch). The input of the amplifier in Fig. 6-5 would then be connected to one vertical input; the output of the amplifier would be connected to the other vertical input. Phase would be checked from the dual pattern by the method explained in Section 5.13.

6.7 Checking Distortion

The oscilloscope may be used in several ways to check harmonic distortion. Some tests are purely qualitative (they serve only to show *presence* of distortion). Others are quantitative; they are concerned not only with presence but with amount of distortion. In either case, the oscilloscope must have low-distortion horizontal and vertical channels. Several tests are described in the following paragraphs.

Sine Wave Patterns. The simplest qualitative test of distortion is made by (a) applying a low-distortion sine-wave test signal to the amplifier; (b) observing the output with an oscilloscope connected to the amplifier output terminals; and (c) noting the deviation of the output waveform from true sinusoidal. If the distortion is appreciable, the disfigurement of the pattern may quickly be spotted by eye. Generator distortion may be observed in the same way when the oscilloscope is connected directly to the generator output.

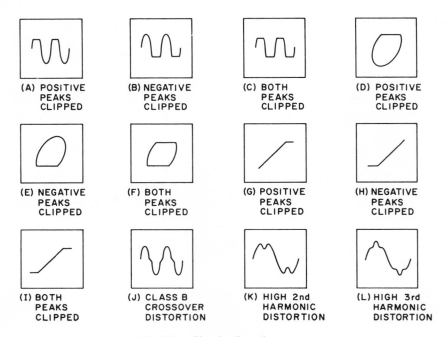

Fig. 6-6. Simple distortion patterns.

Figure 6-6 shows some of the distortion patterns obtained in this type of test. Those in Fig. 6-6A, B, C, J, K, and L are distorted sine waves reproduced with the internal sweep switched on. Those in Fig. 6-6D, E, F, G, H, and I are distorted Lissajous phase-shift figures obtained when the oscilloscope is operated in the manner explained in Section 5.1.

When a dual-trace oscilloscope (or single-trace oscilloscope with electronic switch) is used, the input and output signals of an amplifier under test may be positioned closely on the screen for comparison of waveform. Allowance must be made, however, for any distortion in the electronic switch.

Use of Fundamental Suppressor. Several distortion meters operate by removing the fundamental frequency of a low-distortion test signal from the amplifier output and indicating the amplitude of the remaining voltage. If this latter voltage is due to harmonics produced by distortion in the amplifier, as it is assumed to be, the distortion percentage is indicated by the ratio of the total harmonic voltage to the amplifier output voltage containing fundamental and harmonics.

Several circuits have been designed for this type of measurement. Figure 6-7 shows the basic elements of one of these. Here, the fundamental suppressor is a bridged-T network (L-C_1-C_2-R_s). Inductance L and identical capacitances C_1 and C_2 are chosen for null at the desired test frequency fundamental (f):

$$C_1 = C_2 = 1/(19.75\ f^2\ L)$$

where C is in farads, f in hertz, and L in henrys. The Q of the inductor should be 10 or higher. Adjustment of the rheostat, R_s, deepens the null. The required resistance at null is inversely proportional to the equivalent series resistance of inductor L.

The amplifier is driven by a low-distortion sine-wave signal generator and, if a power amplifier, is terminated with a resistance R_L equal to its load impedance. Switch S is thrown first to position 1, and the entire output voltage (E_1) of the amplifier is measured with the oscilloscope. S then is thrown to position 2, rheostat R_s is adjusted for null, and the harmonic voltage (E_2) is measured. Distortion percentage is then calculated:

$$D(\%) = 100\ (E_2/E_1).$$

Measurement Procedure

1. Set up oscilloscope.
2. Switch on internal sweep.
3. Set SYNC SELECTOR switch to INTERNAL.
4. Set up equipment as shown in Fig. 6-7.
5. Switch on amplifier and set its gain and tone controls to desired operating point.
6. Switch on generator and set it to null frequency of bridged-T network. Adjust generator output for normal power output (or voltage output) of amplifier.
7. Set switch S to position 1.

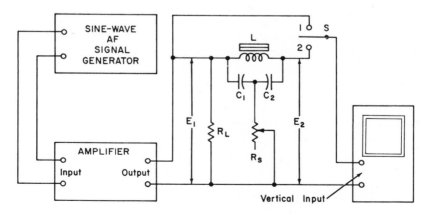

Fig. 6-7. The setup for harmonic distortion measurement.

8. Adjust VERTICAL GAIN control until pattern appears on screen.
9. Adjust SWEEP FREQUENCY, HORIZONTAL GAIN, and SYNC controls for several stationary cycles on screen.
10. From this pattern, measure voltage E_1, using one of methods outlined in Chapter 4.
11. Throw switch S to position 2, adjust R_s for deepest null, and record the voltage at null as E_2.
12. Calculate total harmonic distortion:

$$D(\%) = 100 \, (E_2/E_1)$$

13. Repeat distortion measurement at several settings of amplifier gain control and at several test frequencies (C_1 and C_2 must be changed for a new frequency).

A sensitive, accurately calibrated oscilloscope is superior to a v-t voltmeter in this application, since the oscilloscope shows the waveform as well as the amplitude of the harmonic voltage. The meter can give erroneous indications when it has been calibrated on a sine-wave basis and the distortion voltage is nonsinusoidal.

Oscilloscope with Tunable Distortion Meter. Another fundamental suppression type of distortion meter employs a tunable RC network continuously to vary the null frequency. This avoids the single-frequency inconvenience of the bridged-T filter shown in Fig. 6-7. This type of instrument has a meter reading directly in distortion percentage, but it also has an output jack or terminals for connection of an oscilloscope to monitor the harmonic waveform.

6.8 Checking Intermodulation

To check intermodulation, a mixed signal (consisting of a low and a high frequency) is applied to the input of the amplifier under test. If there is intermodulation in the amplifier, the output signal of the amplifier will be amplitude modulated (the higher-frequency signal modulated by the lower-frequency one). From the AM waveform displayed on the oscilloscope screen, the intermodulation percentage may be determined.

Figure 6-8A shows a test setup for intermodulation measurement. The generator supplies a mixed signal, such as 60 and 7000 Hz. The amplifier is terminated by a resistor R_L that is equal to its normal load impedance. The high-pass filter transmits the modulated high-frequency signal, but blocks the low-frequency signal. The AM wave is displayed on the oscilloscope screen and resembles Fig. 6-8B.

Measurement Procedure

1. Set up oscilloscope.
2. Switch on internal sweep.

Audio Amplifier, Receiver, Transmitter Tests & Measurements 115

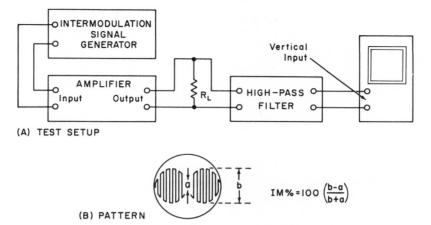

Fig. 6-8. The setup for intermodulation measurement.

3. Set SYNC SELECTOR switch to INTERNAL.
4. Set up equipment as shown in Fig. 6-8A.
5. Switch on amplifier and set its gain and tone controls to desired operating point.
6. Switch on generator and adjust its output for normal power output of amplifier.
7. Adjust VERTICAL GAIN control until pattern appears on screen.
8. Adjust SWEEP FREQUENCY and SYNC controls for several stationary cycles on screen (see Fig. 6-8B).
9. Adjust HORIZONTAL and VERTICAL GAIN controls for desired width and height of pattern.
10. Measure vertical dimensions *a* and *b* (see Fig. 6-8B) in screen divisions.
11. Calculate intermodulation percentage:

$$IM(\%) = 100\,[(b-a)/(b+a)]$$

12. Repeat intermodulation measurement at several settings of gain and tone controls of amplifier.

6.9 Square-Wave Testing

The manner in which an amplifier or component handles a square-wave test signal gives a concise estimate of its performance at a number of sine-wave frequencies. A good square wave is applied to the amplifier input, and the output signal is viewed with an oscilloscope. The degree to which the square wave is deformed in passing through the amplifier reveals certain defects in performance; if squareness is preserved, frequency and phase response of the amplifier are good up to approximately 9 times

116   *Practical Oscilloscope Handbook*

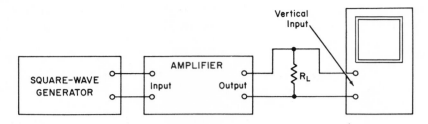

Fig. 6-9. The setup for square-wave testing.

the square-wave frequency. The square wave has high odd-harmonic content. Thus, the 3rd, 5th, 7th, and 9th harmonics are emphasized.

The square-wave generator used in this test should deliver a signal having excellent waveform: flat top, fast rise, fast fall, and negligible overshoot. The oscilloscope channels must themselves have excellent square-wave response. Figure 6-9 shows the test setup.

Figure 6-10 shows some of the output square-wave patterns and their interpretations. It is evident from these that degraded low-frequency

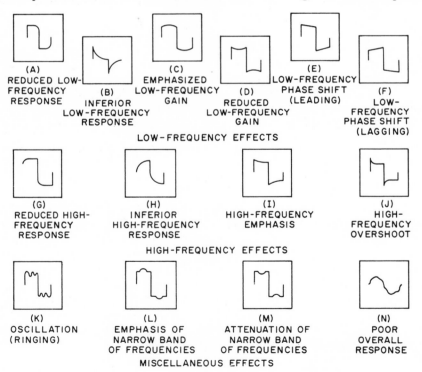

Fig. 6-10. Square-wave test patterns.

Audio Amplifier, Receiver, Transmitter Tests & Measurements 117

response tends to round the leading edge of the wave (Fig. 6-10A), whereas degraded high-frequency response rounds the trailing edge (Fig. 6-10C).

Test Procedure

1. Set up oscilloscope.
2. Switch on internal sweep.
3. Set SYNC SELECTOR switch to INTERNAL.
4. If unit under test is power amplifier, terminate it with load resistance R_L equal to load impedance of amplifier.
5. Set up equipment as shown in Fig. 6-9.
6. Switch on amplifier and set its gain control and tone control to desired operating point.
7. Switch on generator and increase its output for peak-to-peak square-wave amplitude not higher than the peak-to-peak voltage that will overdrive amplifier.
8. Adjust VERTICAL GAIN control until pattern appears on screen.
9. Adjust SWEEP FREQUENCY and SYNC controls for one stationary square-wave cycle on screen.
10. Adjust HORIZONTAL and VERTICAL GAIN controls for desired width and height of pattern.
11. Note squareness of pattern, comparing it with samples given in Fig. 6-10 for possible interpretation.
12. Repeat test at several square-wave frequencies and at several settings of the amplifier gain and tone controls.

6.10 Oscilloscope as AF Signal Tracer

The oscilloscope is superior to the v-t voltmeter as a high-impedance signal tracer in AF troubleshooting since it will not only show the presence and amplitude of the signal at a test point, but will also reveal its waveform. The oscilloscope will also reveal oscillation, hum, and noise, and will indicate distortion of the signal.

In this application, the oscilloscope is employed in the manner of a voltmeter. When tracing a signal, use a low-capacitance probe, and follow either of the voltage-measuring methods outlined previously. Supply a low-distortion, sine-wave test signal to the amplifier (or circuit) input terminals, and move the oscilloscope probe from stage to stage, progressing from input to output, to check presence and amplitude of signal. The amplifier must be in operation and its gain and tone controls set for normal response. A 1,000-Hz signal is preferable in most cases.

6.11 Oscilloscope as Bridge Null Detector

When used as the null detector for an ac bridge, the oscilloscope gives separate indications for reactive balance and resistive balance of the bridge.

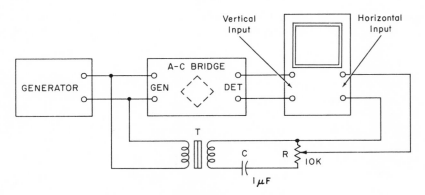

Fig. 6-11. The oscilloscope as a null detector.

Figure 6-11 shows how an oscilloscope is connected to the bridge. Here, T is a shielded transformer, and capacitor C and potentiometer R form an adjustable phase-shift network. The generator voltage is applied simultaneously to the oscilloscope horizontal input (through the phase shifter) and to the bridge input. The bridge output signal is applied to the oscilloscope vertical input. Figure 6-12 shows the type of patterns obtained.

Test Procedure

1. Set up oscilloscope.
2. Switch off internal sweep.
3. Set SYNC SELECTOR switch to EXTERNAL.
4. Set HORIZONTAL and VERTICAL GAIN controls to midrange.
5. Connect equipment as shown in Fig. 6-11.
6. With bridge unbalanced but with test component (resistor, capacitor, inductor) connected to bridge UNKNOWN terminals, adjust R in phase shifter to give ellipse pattern on screen.
7. Readjust HORIZONTAL and VERTICAL GAIN controls for ellipse of suitable size.
8. Adjust reactance control of bridge, noting that ellipse tilts to right (Fig. 6-12A) or to left (Fig. 6-12C). When reactance balance is complete (reactance null), ellipse will be horizontal (Fig. 6-12B).
9. Adjust resistance (power factor, dissipation factor, or Q) control of bridge, noting that ellipse closes.
10. At complete null (reactance and resistance both balanced), straight horizontal line is obtained (Fig. 6-12E). If resistance is balanced while reactance is unbalanced, tilted ellipse will close, giving single-line trace tilted to the right (Fig. 6-12D) or to the left (Fig. 6-12F).

A simpler but less effective way to use an oscilloscope as a null detector is to connect the bridge output (DETECTOR terminals) to the

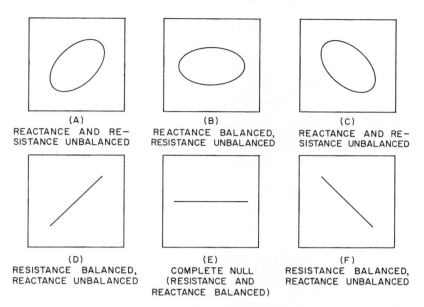

Fig. 6-12. Null detector patterns.

vertical input and use the oscilloscope as a voltmeter which gives its lowest reading at null. This method may be used with internal sweep either on or off.

6.12 Distortion Measurement with a Dual-Trace Scope

An amplifier stage, or an entire amplifier unit, may be tested for distortion with a dual-trace oscilloscope. This type of measurement is especially valuable when the slope of a waveform must be faithfully reproduced by an amplifier. Fig. 6-13 shows the testing of such a circuit using a triangular wave, such as is typically encountered in the recovered audio output of a limiting circuit which precedes the modulator of a transmitter. The measurement may be made using any type of signal; merely use the waveform for testing that is normally applied to the amplifier during normal operation.

Measurement Procedure

1. Apply the type of signal normally encountered in the amplifier under test.
2. Connect Channel A probe to the input of the amplifier and Channel B probe to the output of the amplifier. It is preferable if the two signals are not inverted in relationship to each other, but inverted signals can be used.
3. Set CH A and CH B DC-GND-AC switches to AC.

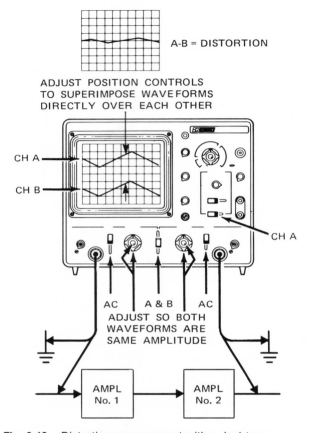

Fig. 6-13. Distortion measurement with a dual-trace scope.

4. Set MODE switch to A & B.
5. Set sync SLOPE switch to CH A and adjust controls for synchronized waveforms.
6. Adjust the CH A and CH B POSITION controls to superimpose the waveforms directly over each other.
7. Adjust the CH A and CH B vertical sensitivity controls (VOLTS/CM and VARIABLE) so that the waveforms are as large as possible without exceeding the limits of the scale, and so that both waveforms are exactly the same height.
8. Now set the MODE switch to the A-B position (if one waveform is inverted in relationship to the other, use the A+B position). Adjust the fine vertical sensitivity (CH B VARIABLE) slightly for the minimum remaining waveform. If the two waveforms are exactly the same amplitude and there is no distortion, the waveforms will cancel

Audio Amplifier, Receiver, Transmitter Tests & Measurements

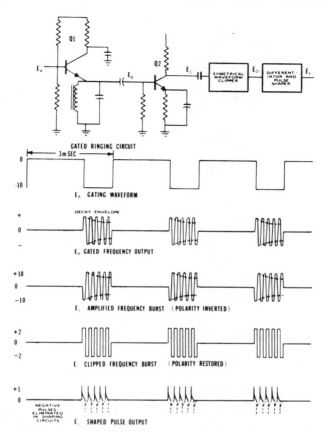

Fig. 6-14. Gated ringing circuit and waveforms. *(Courtesy* Dynascan Corp.)

and there will be only a straight horizontal line remaining on the screen. Any waveform that remains represents the distortion existing.

6.13 Checking Gated Ringing Circuit with Dual-Trace Scope

The circuit and waveforms of Fig. 6-14 are shown to demonstrate the type of circuit in which the dual-trace oscilloscope is effective both in design and in troubleshooting applications. Waveform A is the reference waveform and is applied to Channel A input. All other waveforms are sampled at Channel B and compared to the reference waveform of Channel A. The frequency burst signal can be examined more closely either by increasing the sweep time per centimeter to 0.5 msec per centimeter or by selecting the 5 times magnification feature. This control can then be rotated to center the desired waveform information on the oscilloscope screen.

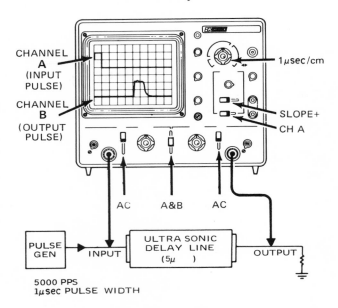

Fig. 6-15. Delay line measurements. *(Courtesy* Dynascan Corp.*)*

6.14 Delay Line Tests with a Dual-Trace Scope

A dual-trace oscilloscope can also be used to determine the delay times of transmission-type as well as ultrasonic-type delay lines. The input pulse can be used to trigger or synchronize the Channel A display and the delay line output can be observed on Channel B. A repetitive-type pulse will make it possible to synchronize the displays. The interval between repetitive pulses should be large compared to the delay time to be investigated. In addition to determining delay time, the pulse distortion inherent in the delay line can be determined by examination of the delayed pulse observed on the Channel B waveform display. Figure 6-15 demonstrates the typical oscilloscope settings as well as the basic test circuit. Typical input and output waveforms are shown on the oscilloscope display. Any pulse stretching and ripple can be observed and evaluated. The results of modifying the input and output terminations can be observed directly.

A common application of the delay line checks is found in color television receivers. Figure 6-16 shows the oscilloscope settings and typical circuit connections to check the "Y" delay line employed in the video amplifier section. The input waveform and the output waveform are compared for delay time, using the horizontal sync pulse of the composite video signal for reference. The indicated delay is approximately one

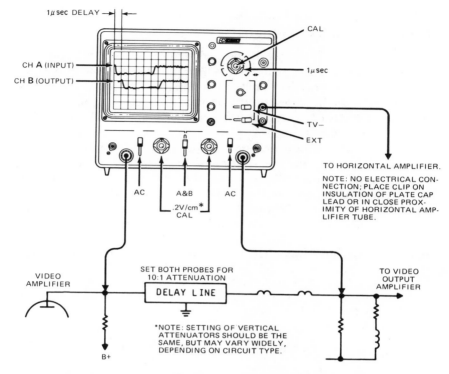

Fig. 6-16. Checking Y-delay line in a color TV receiver. *(Courtesy Dynascan Corp.)*

microsecond. In addition to determining the delay characteristics of the line, the output waveform reveals any distortion that may be introduced from an impedance mismatch or a greatly attenuated output resulting from an open delay line.

6.15 Stereo Amplifier Servicing with a Dual-Trace Scope

Another convenient use for dual-channel oscilloscopes is in troubleshooting stereo amplifiers. If identical channel amplifiers are used and the output of one is weak, distorted, or otherwise abnormal, the dual-trace oscilloscope can be efficiently used to localize the defective stage. With an identical signal applied to the inputs of both amplifiers, a side-by-side comparison of both units can be made by progressively sampling identical signal points in both amplifiers. When the defective or malfunctioning stage has been located, the effects of whatever troubleshooting and repair methods are employed can be observed and analyzed immediately.

6.16 Using a Dual-Trace Scope to Improve the Ratio of Desired-to-Undesired Signals

In some applications, the desired signal may be riding on a large undesired signal component, such as 60 Hz. It is possible to minimize or, for practical purposes, eliminate the undesired component. Figure 6-17 indicates the oscilloscope control settings for such an application. The waveform display of Channel A indicates the desired signal, and the dotted line indicates the average amplitude variation corresponding to an undesired 60-Hz component. The Channel B display indicates a waveform of equal amplitude and identical phase to the average of the Channel 1 waveform. With the MODE switch set to A − B, and by adjusting the CH B vertical attenuator controls, the 60-Hz component of the Channel A signal can be canceled by the Channel B input and the desired waveform can be observed without the 60-Hz component.

6.17 Amplifier Phase Shift Measurements with Dual-Trace Scope

In all amplifiers, a phase shift is always associated with a change in amplitude response. For example, at the −3 dB response points, a phase

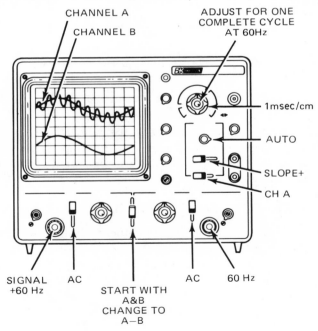

Fig. 6-17. Improving desired-to-undesired signal ratio. *(Courtesy* Dynascan Corp.)

Audio Amplifier, Receiver, Transmitter Tests & Measurements 125

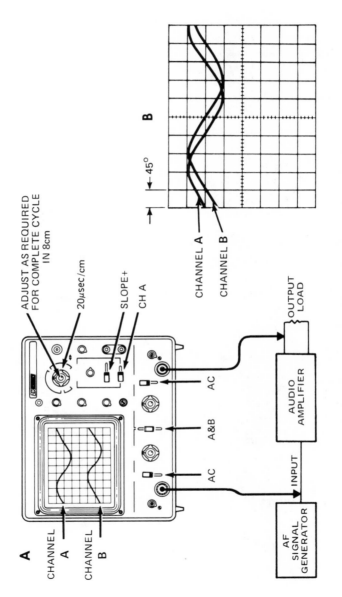

Fig. 6-18. Measuring amplifier phase shift with dual-trace scope.

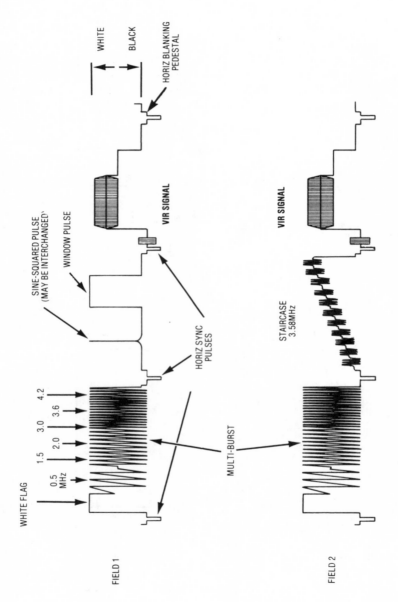

Fig. 6-19. VITS signal, fields 1 and 2. (Courtesy Dyanscan Corp.)

shift of 45° occurs. Figure 6-18A illustrates a method of determining amplifier phase shift directly. In this particular case, the measurements are being made at approximately 5,000 Hz. The input signal to the audio amplifier is used as a reference and is applied to the CH A INPUT jack.

The SWEEP VARIABLE control is adjusted as required to provide a complete cycle of the input waveform displayed on 8 cm horizontally. A waveform height of 2 cm is used. The 8-cm display represents 360° at the displayed frequency, and each centimeter represents 45° of the waveform. The signal developed across the output of the audio amplifier is applied to the Channel B INPUT jack. The vertical attenuator controls of Channel B are adjusted as required to produce a peak-to-peak waveform of 2 cm as shown in Fig. 6-18B.

The CH B POSITION control is then adjusted so that the Channel B waveform is displayed on the same horizontal axis as the Channel A waveform, as shown in Fig. 6-18B. The distance between corresponding points on the horizontal axis for the two waveforms then represents the phase shift between the two waveforms. In this case, the zero crossover points of the two waveforms are compared. It is shown that a difference of 1 cm exists. This is then interpreted as a phase shift of 45°.

6.18 Checking VITS (Vertical Interval Test Signal) with a Dual-Trace Scope

Many television servicing procedures can be performed using single-trace operation. One of these procedures, viewing the VITS (vertical interval test signal) can be accomplished much more effectively by using a dual-trace oscilloscope. First, let's analyze VITS and single-trace testing.

Most network television signals contain a built-in test signal (the VITS) that can be a very valuable tool in troubleshooting and servicing television sets. This VITS can localize trouble to the antenna, tuner, IF, or video sections and shows when realignment may be required. The following procedures show how to analyze and interpret oscilloscope displays of the VITS.

The VITS is transmitted during the vertical blanking interval. It can be seen as a bright white line above the top of the picture when the vertical linearity or height is adjusted to view the vertical blanking circuits (the blanking circuit must be disabled to see the VITS).

The transmitted VITS is a precision sequence of specific frequency, amplitude, and waveshape as shown in Figs. 6-19 and 6-20. The television networks use the precision signals for adjustment and checking of network transmission equipment, but the technician can use them to evaluate television set performance. The first frame of the VITS (line 17) begins with a "flag" of white video, followed by sine-wave frequencies of 0.5 MHz, 1.5 MHz, 2 MHz, 3 MHz, 3.6 MHz (3.58 MHz), and 4.2 MHz. This sequence of frequencies is called the "multi-burst." The first frame of Field #2 (line 279) also contains an identical multi-burst. This multi-burst portion of the

128   *Practical Oscilloscope Handbook*

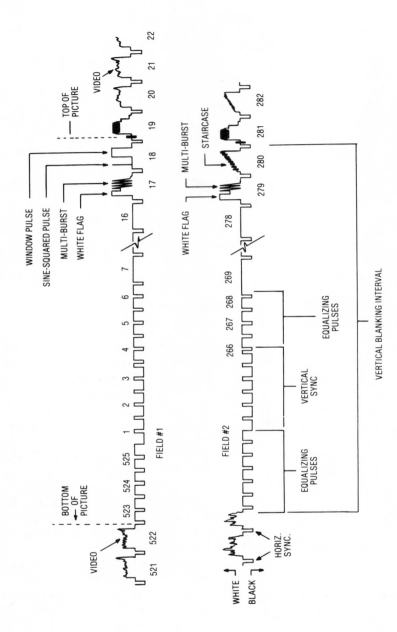

Fig. 6-20. Vertical blanking interval, showing VITS information. *(Courtesy Dynascan Corp.)*

Audio Amplifier, Receiver, Transmitter Tests & Measurements 129

VITS is the portion that can be most valuable to the technician. The second frame of the VITS (lines 18 and 280), which contains the sine-squared pulse, window pulse, and the staircase of 3.58-MHz bursts at progressively lighter shading, is valuable to the network, but has less value to the technician. As seen on the television screen, Field #1 is interlaced with Field #2 so that line 17 is followed by line 279 and line 18 is followed by line 280. The entire VITS appears at the bottom of the vertical blanking pulse and just before the first line of video.

Each of the multi-burst frequencies is transmitted at equal strength. By observing the comparative strengths of these frequencies after the signal is processed through the television receiver, the frequency response of the set may be checked.

Set up the oscilloscope as follows to view the VITS (refer to Section 2.26):

Measurement Procedure for Single-Trace

1. Connect the CH B probe (set at 10:1) to the output of the video detector.
2. If the television set has a vertical retrace blanking circuit, bypass this circuit during the measurement.
3. Set the MODE switch to CH B.
4. Set up the oscilloscope for TV vertical composite video waveform analysis. Two vertical frames will be visible.
5. Place the sweep time VARIABLE control in the CAL position.
6. Reduce sweep time to 0.1 msec per cm (0.1 msec/cm) with the SWEEP TIME/CM switch. This expands the display by increasing the sweep speed. The VITS information will appear to the right on the expanded waveform display.
7. Further expand the sweep with the 5X magnification control. Rotate the POSITION control in a counterclockwise direction, moving the trace to the left, until the expanded VITS appears. The brightness level of the signal display will be reduced due to the use of 5X magnification.
8. The waveform should be similar to that shown in Fig. 6-21. For the oscilloscope display, each vertical sync pulse starts a new sweep. This causes line 17 and line 279 to be superimposed, as are lines 18 and 280. The multi-burst signals are identical, which reinforces the trace. However, lines 18 and 280 are not identical, and both signals are superimposed over each other.
9. The above steps represent the limit of observation possible with a single-trace oscilloscope. A single-field VITS presentation can be obtained by placing the MODE switch in the A & B position. This causes the Channel B information to be displayed on alternate sweeps, as are the Field #1 and Field #2 VITS. Because there is no provision for preselecting the Field #1 or Field #2 information, either

130 *Practical Oscilloscope Handbook*

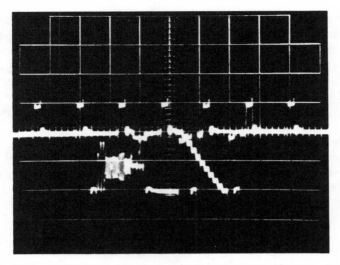

Fig. 6-21. VITS signal viewed on a single-trace scope. *(Courtesy* Dyanscan Corp.)

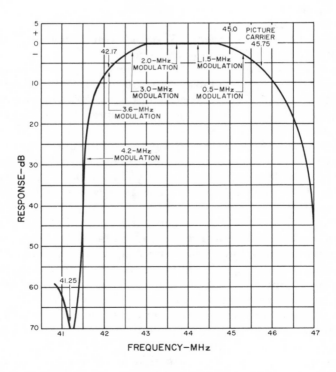

Fig. 6-22. Color TV IF amplifier response curve. *(Courtesy* Dynascan Corp.)

Field #1 or Field #2 will appear. The multi-burst information in the VITS is the most valuable for troubleshooting television receivers because it is present on both Field #1 and Field #2.

Now to analyze the waveform. All frequencies of the multi-burst are transmitted at the same level but should not be equally coupled through the receiver because of its response curve. Figure 6-22 shows the desired response for a good color television receiver, identifying each frequency of the multi-burst and showing the allowable amount of attentuation for each. Remember that –6 dB equals half the reference voltage (the 2.0-MHz modulation should be used for reference).

To identify receiver problems, start by observing the VITS at the video detector. This will localize trouble to a point either before or after the detector. If the multi-burst is normal at the detector, check the VITS on other channels. If some channels appear normal but others do not, tuner or attenna-system troubles are likely. Don't overlook the possibility of the antenna system causing "holes" or tilted response on some channels. If the VITS is abnormal at the video detector on all channels, the trouble is probably in the IF amplifier stages.

As another example, let us assume that we have a set on the bench with a poor-quality picture. The oscilloscope shows the VITS at the video detector to be about normal except that the burst at 2.0 MHz is low compared to the bursts on either side. This suggests an IF trap is detuned into the passband, removing frequencies 2 MHz below the picture carrier frequency. Switch to another channel carrying VITS. If the same display is seen, then the analysis is correct, and the IF amplifier requires realignment. If the poor response at 2 MHz is not seen on other channels, perhaps an FM trap at the tuner input is misadjusted, causing a dropoff on only one channel. Other traps at the input of the TV set could be similarly misadjusted or faulty.

Finally, if the VITS response at the detector output is normal for all channels, the trouble will be in the video amplifier.

As shown in Figs. 6-19 and 6-20, the information on the Field #1 and Field #2 vertical blanking interval pulse is different. Also, because the oscilloscope sweep is synchronized to the vertical blanking interval waveform, the Field #1 and Field #2 waveforms are superimposed onto each other as shown in Fig. 6-21. With dual-trace operation, the signal information on each blanking pulse can be viewed separately without overlapping. Figure 6-23 indicates the oscilloscope control setting for viewing the alternate VITS.

Measurement Procedure for Dual-Trace

1. The color television receiver on which the VITS information is to be viewed must be set to a station transmitting a color broadcast.
2. The control settings of Fig. 6-23 are those required to obtain a two-field vertical display on Channel A.

132   *Practical Oscilloscope Handbook*

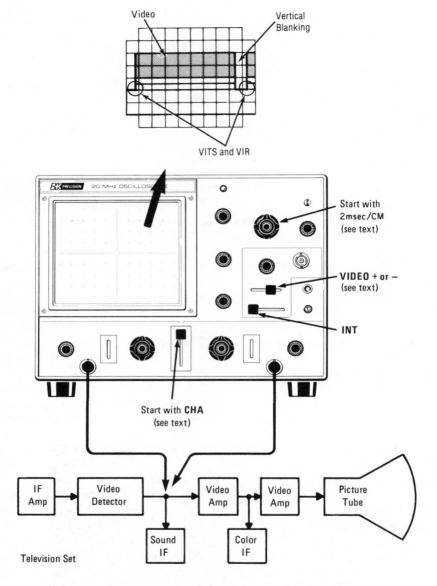

Fig. 6-23. Setup to view VITS signal on dual-trace scope. *(Courtesy Dynascan Corp.)*

3. With the oscilloscope and television receiver operating, connect the Channel A probe (set at 10:1) to the video detector test point.
4. Set the SYNC switch as follows:

a. If the sync and blanking pulses of the observed video signal are positive, use the TV+ switch position.
b. If the sync and blanking pulses are negative, use the TV- switch position.
5. Adjust the sweep time VARIABLE control so that two vertical fields are displayed on the oscilloscope screen.
6. Connect the Channel B probe (set to 10:1) to the video detector test point.
7. Set the MODE switch to the A & B position. Identical waveform displays should now be obtained on Channels A and B.
8. Place the sweep time VARIABLE control in the CAL position.
9. Set the SWEEP TIME/CM control to the 0.1 msec/CM position. This expands the display by increasing the sweep speed. The VITS information will appear toward the right-hand portion of the expanded waveform displays. The waveform information on each trace may appear as shown in Fig. 6-20. Because there is no provision for synchronizing the oscilloscope display to either of the two fields that comprise a complete vertical frame, it cannot be predicted which field display will appear on the Channel A or Channel B display.
10. Turn on the 5X magnification control. Rotate the control to move the traces to the left until the expanded VITS information appears as shown in Fig. 6-24. Because of the low repetition rate and the high-sweep speed combination, the brightness level of the signal displays will be reduced.
11. Once the Channel A and Channel B displays have been identified as being either Field #1 or Field #2 VITS information, the probe corresponding to the waveform display that is to be used for signal-

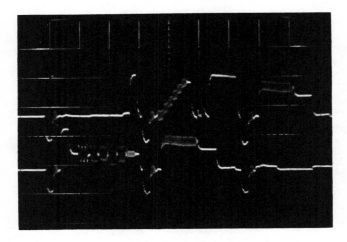

Fig. 6-24. Dual-trace display of VITS lines 1 and 2. *(Courtesy* Dynascan Corp.)

tracing and troubleshooting can be used, and the remaining probe should be left at the video detector test point to ensure that the sync signal is not interrupted. If the sync signal is interrupted, the waveform displays may reverse because, as previously explained, there is no provision in the oscilloscope to identify either of the two vertical fields that comprise a complete frame.

6.19 Visual Alignment of AM IF Amplifier

Figure 6-25A shows the test setup for alignment of the IF channel of an AM receiver. The sweep generator supplies the intermediate frequency (usually 455 kHz) and sweeps this signal by a selected amount above and below center frequency. The sweep generator also supplies a 60 Hz sweep signal for the oscilloscope. The marker generator places a pip on the oscilloscope display for frequency identification.

Test Procedure
1. Set up oscilloscope, with low-capacitance probe.
2. Switch off internal sweep.
3. Set SYNC SELECTOR switch to EXTERNAL.
4. Set up equipment as shown in Fig. 6-25A:
   a. Connect sweep generator to input of IF amplifier.
   b. External marker generator will not be required if sweep generator has internal variable-frequency marker.
   c. Connect VERTICAL INPUT (probe) to 2nd detector load resistor, R, in receiver.
   d. Connect HORIZONTAL INPUT terminals to 60-Hz output of sweep generator.
5. Switch on receiver and generators, reduce output of generators temporarily to zero, and detune receiver from any strong station.

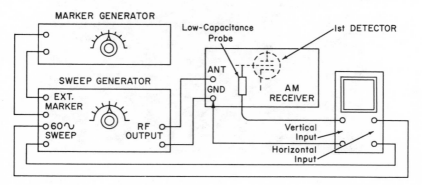

Fig. 6-25. Visual AM IF alignment.

Audio Amplifier, Receiver, Transmitter Tests & Measurements 135

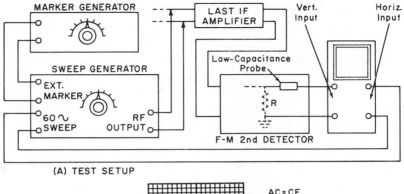

(A) TEST SETUP

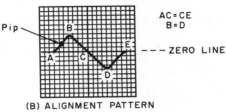

(B) ALIGNMENT PATTERN

Fig. 6-26. FM detector alignment.

6. Set sweep generator output to trial level.
7. Tune sweep generator to intermediate frequency (generally 455 kHz).
8. Adjust sweep to 30 kHz.
9. Adjust VERTICAL GAIN control until pattern appears on screen.
10. Increase sweep generator output, if necessary, to give pattern resembling Fig. 6-25B.
11. Adjust HORIZONTAL and VERTICAL GAIN controls for desired width and height of pattern.
12. If pattern is upside-down, adjust PHASING control in sweep generator to right it.
13. Adjust IF trimmers for narrow, single-line pattern of type shown in Fig. 6-25B. If sweep generator output is excessive, feet of pattern may tilt upward, as in Fig. 6-25C, or pattern may become double-peaked, as in Fig. 6-25E.
14. When trimmer adjustment is staggered, a broad curve (Fig. 6-25D) is obtained; severe staggering gives double-peak curve (Fig. 6-25E).
15. To check frequency points on curve, advance marker generator output. Note that a pip is produced on curve. Keep marker generator output low, or oversized pip will distort response curve. Tune marker generator, noting that pip moves along curve. The frequency at any desired point on curve, marked by pip, may be read from marker generator dial. In this way, bandwidth of IF channel may be determined, or trimmers may be set for a desired bandwidth (e.g., flat-top response for high-fidelity service).

6.20 Visual Alignment of FM Detector

Figure 6-26 shows the test setup for alignment of an FM 2nd detector (discriminator or ratio detector). The sweep generator supplies the intermediate frequency (usually 10.7 MHz) as the center frequency, and sweeps this signal above and below center by a selected number of kilohertz. The sweep generator also supplies a 60-Hz sweep signal for the oscilloscope. The marker generator places a pip on the display for identification of frequency.

The Test Procedure is similar to the AM IF Amplifier procedure, Section 6.19, except the sweep generator is set to 10.7 MHz and sweep width is 300 kHz. Also, the trimmers or coils are adjusted to produce the pattern shown in Fig. 6-26. Peaks B and D should be equal in height, and points A, C, and E should lie along the zero base line.

6.21 Visual Alignment of FM IF Amplifier

The Test Procedure is similar to the AM IF Amplifier procedure, Section 6.19, except the sweep generator is set to 10.7 MHz and sweep width is 300 kHz. The test setup is shown in Fig. 6-27.

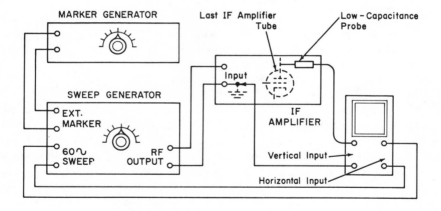

Fig. 6-27. The setup for FM IF alignment.

6.22 Visual Alignment of TV IF Amplifier

For alignment of the video IF amplifier of a black-and-white television receiver, use the test setup shown in Fig. 6-28. Connect the sweep generator to the video IF amplifier input (grid of mixer tube or primary of first IF transformer), and connect the VERTICAL INPUT (probe) to the video IF amplifier output (grid of video amplifier).

The Test Procedure is similar to the AM IF Amplifier procedure, Section 6.19, except that the sweep generator is set to 43 MHz and sweep width is 6

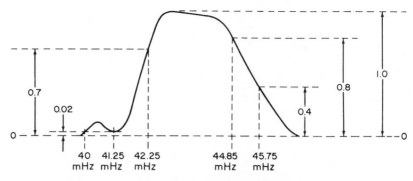

Fig. 6-28. A typical video IF response pattern.

MHz. Adjust trimmers or coils as recommended in the TV receiver manufacturer's service notes to obtain curve similar to that in Fig. 6-28.

6.23 Visual Alignment of TV Sound IF Amplifier and Detector

Follow the Test Procedure for AM IF Amplifier (Section 6.19) and FM DETECTOR (Section 6.20) but set the sweep generator for 4.5 MHz and sweep width to 30 kHz.

6.24 Checking the Video Amplifier with Square Waves

The video amplifier is a wideband unit (20 Hz to 4.5 MHz). Its performance may be rapidly appraised by means of square-wave test signals. For this purpose, the square-wave generator must be tunable from 20 Hz to 500 kHz, and the oscilloscope must have vertical response to 5 MHz or higher and sweep frequency tunable to 500 kHz. The test setup is shown in Fig. 6-29.

Test Procedure

1. Set up oscilloscope with low-capacitance probe.
2. Switch on internal sweep.
3. Set SNYC SELECTOR switch to INTERNAL.
4. Set up equipment as shown in Fig. 6-29:
   a. Connect square-wave generator to video amplifier input.
   b. Connect VERTICAL INPUT (probe) to video amplifier output.
5. Switch on receiver and generator, tune receiver to unused TV channel, and reduce generator output temporarily to zero.
6. Set generator output to trial level below overload level of video amplifier (see manufacturer's literature).

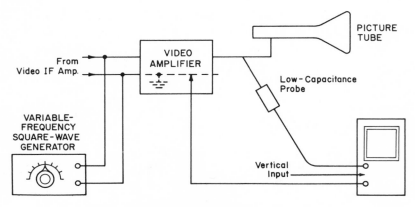

Fig. 6-29. The setup for checking the video amplifier.

7. Tune generator to 20 Hz.
8. Adjust VERTICAL GAIN control until pattern appears on screen.
9. Adjust SWEEP FREQUENCY and SYNC controls for single, stationary square-wave cycle on screen. At high frequencies, beyond the maximum sweep frequency, several cycles will appear.
10. Adjust HORIZONTAL and VERTICAL GAIN controls for desired width and height of pattern.
11. Observe pattern for squareness. Compare with sample patterns given in Fig. 8-10.
12. Repeat test at square-wave frequencies of 50 Hz, 25 kHz, 100 kHz, and 500 kHz.
13. If CONTRAST control of receiver is in video amplifier, repeat complete test procedure at several settings of this control.

6.25 Visual Alignment of TV Front End

For adjustment of the front end of a TV receiver:

1. Tune receiver to desired TV channel frequency.
2. Tune sweep generator to center frequency of that channel.
3. Set sweep to 10 MHz.
4. Tune marker first to video carrier and then to sound carrier of channel to which receiver is tuned, and adjust RF and mixer trimmers for curve of desired bandwidth (as specified by set manufacturer's literature), using marker pips for guidance.

6.26 Checking TV Operating Waveforms

Video and sync pulse voltages and their operating waveforms should be checked visually for shape. At the same time, their peak amplitudes may be measured on the oscilloscope screen.

Audio Amplifier, Receiver, Transmitter Tests & Measurements

For this test, tune in a TV station, and touch the oscilloscope probe successively to each point of interest in the receiver circuit. Observe the resulting pattern for shape, measure its peak voltage, and compare these data with the set manufacturer's specifications. For convenience, some oscilloscopes designed for TV work have, in addition to continuously variable sawtooth sweep, two preset sweep frequencies: 7875 Hz (which will display two cycles at the TV horizontal frequency) and 30 Hz (which will display two cycles at the TV vertical frequency). Compare shape of signal and peak-to-peak voltages with manufacturer's service notes.

6.27 Checking Amplitude Modulation by Sine-Wave Method

The adjustment and troubleshooting of a radio transmitter are facilitated by use of a suitable oscilloscope. In some applications, such as modulation checking, the oscilloscope can show more about the nature of the signal and the condition of the transmitter than can an equivalent test meter. In other applications, the oscilloscope is a useful adjunct to listening tests and to meters inside and outside the transmitter.

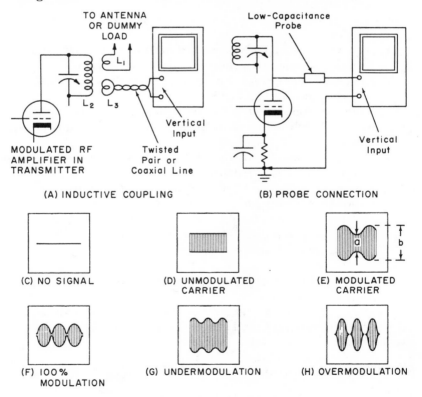

Fig. 6-30. AM checking with sine-wave patterns.

The AM wave is displayed on the screen, and the modulation percentage is determined from the vertical dimensions of the pattern. Figures 6-30A and B show the test setup.

Test Procedure

1. Set up oscilloscope.
2. Switch on internal sweep.
3. Set SYNC SELECTOR switch to INTERNAL.
4. Set up equipment as shown in Fig. 6-30A or B:
   a. If frequency response of vertical amplifier does not reach transmitter carrier frequency, use direct input to vertical deflecting plates through a small 3-turn pickup coil (L_3) coupled to tank of modulated RF amplifier in transmitter, as shown in Fig. 6-30A. If oscilloscope can handle carrier frequency, vertical amplifier may be used with either inductive coupling (Fig. 6-30A) or low-capacitance probe (Fig. 6-30B). With latter method, combined dc and RF voltage in transmitter must not exceed breakdown voltage of probe and oscilloscope.
   b. With transmitter OFF, set VERTICAL GAIN control to mid position.
   c. Set SWEEP FREQUENCY and HORIZONTAL GAIN controls for a single-line trace over most of screen (Fig. 6-30C).
   d. Switch on *unmodulated* transmitter. Unmodulated carrier gives pattern shown in Fig. 6-30D.
   e. Modulate transmitter with low-distortion, sine-wave audio frequency. Amplitude modulation gives wave pattern similar to Fig. 6-30E, F, G, or H.
   f. Adjust HORIZONTAL and VERTICAL GAIN controls for desired width and height of pattern.
   g. Adjust SWEEP FREQUENCY and SYNC controls for 2 or 3 stationary cycles of pattern on screen.
   h. Measure minimum height a and maximum height b of pattern in number of screen divisions, as shown in Fig. 6-30E.
   i. Calculate modulation percentage:

$$M(\%) = 100[(b-a)/(b+a)]$$

Figure 6-30F shows shape of pattern for a fully modulated wave (M = 100%). Figures 6-30G and H show shapes for undermodulation and overmodulation, respectively.

6.28 Checking Amplitude Modulation with Trapezoidal Patterns

This method is somewhat simpler than the one described in Section 6.27 because it does not involve use of the internal sweep of the

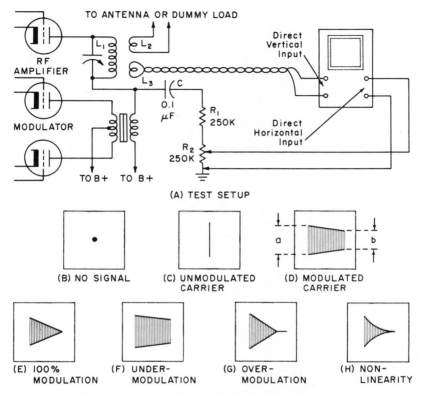

Fig. 6-31. AM checking with trapezoidal patterns.

oscilloscope. It is widely used for direct checking of amplitude modulation and for the calibration of modulation meters. Figure 6-31A shows the test setup.

Test Procedure

1. Set up oscilloscope.
2. Switch off internal sweep.
3. Set SYNC SELECTOR switch to EXTERNAL.
4. Set up equipment as shown in Fig. 6-31A:
   a. Do not use oscilloscope amplifiers. Couple direct vertical input to tank of modulated RF amplifier in transmitter through small 3-turn pickup coil L_3. Connect direct horizontal input to output of modulator through voltage divider network C-R_1-R_2.
   b. With transmitter OFF, set HORIZONTAL and VERTICAL GAIN controls to mid position. Single-dot pattern (Fig. 6-31B) appears on screen.
   c. Switch on *unmodulated* transmitter. Unmodulated carrier gives pattern shown in Fig. 6-31C.

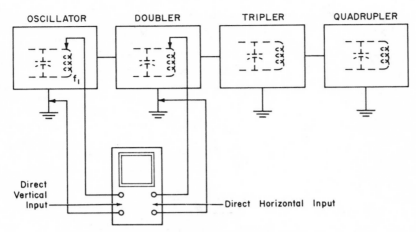

Fig. 6-32. Checking frequency-multiplier operation.

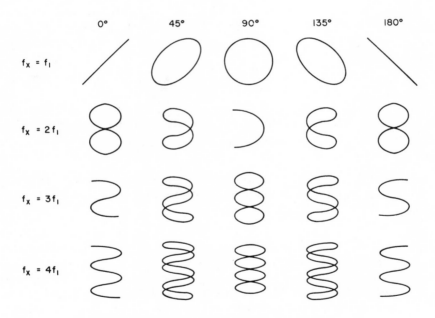

Fig. 6-33. Lissajous figures for transmitter testing.

d. Modulate transmitter with low-distortion, sine-wave audio frequency. Amplitude modulation gives trapezoidal pattern similar to Figs. 6-31D, E, F, G, or H.

e. Adjust coupling between L_1 and L_3 for height, and R_2 for width, to spread pattern over useful area of screen.

f. Measure maximum height *a* and minimum height *b* of pattern in number of screen divisions, as shown in Fig. 6-31D.

g. Calculate modulation percentage:

$$M(\%) = 100[(a-b)/(a+b)]$$

Figure 6-31E shows the trapezoid shape for a fully modulated wave (M = 100%). Figures 6-31F and G show shapes for undermodulation and overmodulation, respectively. An advantage of the trapezoidal pattern is its ability to show linearity of the modulated RF amplifier; the straightness of the sides of the trapezoid indicates this linearity. In Figure 6-31H, the sides are curved, indicating nonlinearity.

6.29 Checking Modulator Channel

The speech amplifier-modulator channel of a transmitter is an AF system. Its performance may be checked in terms of gain, frequency response, harmonic distortion, phase shift, intermodulation, power output, hum, and noise.

For explanations of how to use the oscilloscope to make these tests, refer to Sections 6.1 to 6.9.

6.30 Checking Frequency-Multiplier Operation

In the adjustment of a transmitter, especially a new one, there is a great possibility that an RF amplifier, doubler, tripler, or quadrupler may be tuned to the wrong frequency. This can have serious consequences (legal as well as technical). Because of harmonic response, frequency meters sometimes confuse the operator, leaving him unsure of transmitter adjustments.

Figure 6-32 shows a test setup for using Lissajous figures to determine frequency (f_x) of an amplifier or multiplier stage with respect to the frequency (f_1) of the oscillator stage. In this application, direct input to the horizontal and vertical deflecting plates via the shortest possible leads must be used. Figure 6-33 shows the Lissajous figures obtained for $f_x:f_1$ ratios of 1:1, 2:1, 3:1, and 4:1 (straight through, doubling, tripling, and quadrupling) at phase difference angles of 0°, 45°, 90°, 135°, and 180°.

Test Procedure

1. Set up oscilloscope.
2. Switch off internal sweep.
3. Set SYNC SELECTOR switch to EXTERNAL.
4. Set up equipment as shown in Fig. 6-32.
   a. Using shortest possible leads, connect direct vertical input to

oscillator tank circuit, and connect direct horizontal input to tank circuit of stage under inspection.
b. Set **HORIZONTAL** and **VERTICAL GAIN** controls to zero.
c. Switch on transmitter.
d. Compare with Fig. 6-33 to determine if stage under inspection is passing oscillator frequency or is doubling, tripling, or quadrupling that frequency.

7
servicing the oscilloscope

The oscilloscope is the most convenient tool for troubleshooting electronic circuits by means of waveform observation and analysis. Obviously, then, the technician must place complete confidence in the proper performance of his scope.

When he observes a nonlinear sawtooth waveform at the grid of a TV horizontal amplifier, as an example, he will proceed with troubleshooting the deflection system on the basis that a defective component is the culprit. There should be little doubt in his mind that the scope waveform is a true display of the waveform under examination; however, what if the TV receiver waveform is actually correct but the scope vertical amplifier has developed a problem? The technician may waste considerable time and face deep frustrations as he attempts to correct a proper waveform that his scope display indicates is improper. The greatest loss in such an instance is the eventual realization by the technician that the scope is at fault; his confidence in the future use of this tool may be seriously affected.

Although there is no magical device to keep a scope, or any other electronic device, free from failure, a regularly scheduled maintenance procedure will help considerably in keeping the scope in top condition.

7.1 Keep the Instruction Manual Handy

When a scope is purchased, the instrument is generally supplied with a thoroughly detailed instruction manual. The first rule: don't misplace it. Place it in an envelope and tape it to the side of the scope, where it won't get lost. Or file it where you can find it quickly. The instruction manual should be read carefully, several times. Make notes along the margin to remind you of key points; underline sentences you feel are significant to the uses you have in mind. When you find yourself involved in new applications and scope waveforms do not seem proper, go back to the manual and check whether the setup adjustments are correct.

Many instances in which scope displays appear improper have been traced to incorrect setup or control adjustments. So before coming to a hasty conclusion that your scope needs repair or recalibration, refer to the instruction manual to review the functions of the various operating controls.

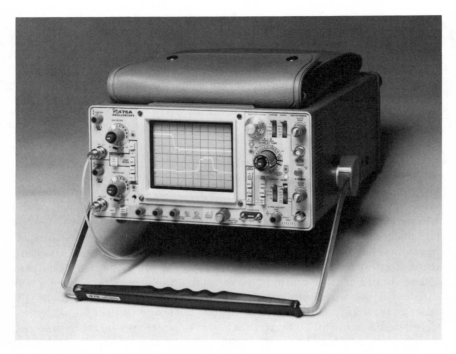

Fig. 7-A. A handy pouch holds the scope's instruction manual and accessory probes. *(Courtesy* Tektronix.)

When working with high-frequency circuits, you will often discover that faulty ground connections pose a serious problem. The display will change in appearance if the test leads are moved or if the case of the scope is touched. At high frequencies, stray capacitances can cause such symptoms. Make sure all ground connections are firm and that ground clips are clean. If necessary, place all equipment on a metal plate to reduce grounding effects; of course, be sure that all equipment or devices under test are isolated from the power line. Check that power transformers are used in the various equipment to be placed on the common metal ground plate; use isolation transformers for each device that does not contain its own power transformer. When proper grounding is completed, the observed display should not change if the test leads are moved or if the scope or other involved instrument cases are touched. One further reminder; use the shortest possible lead lengths when working with high-frequency circuits to reduce stray capacitance effects.

7.2 Contact the Manufacturer

Today's scopes are complex and sophisticated in their design, with the ability to perform more functions with greater conveniences than scopes sold a decade or so ago. Almost all scopes sold recently are solid-state, including integrated circuits (ICs). Thus, when a scope does require service, a certain degree of circuit and troubleshooting knowledge is important; a sure way to multiply a simple scope defect is to plunge into the instrument and fiddle with factory-adjusted controls. If you consider yourself capable of reading and understanding the setup, calibration, and maintenance instructions included in most scope instruction manuals, you are ready to proceed to service the scope. Otherwise, contact the manufacturer for an estimate and learn where to deliver or ship the defective scope. It is not uncommon to have your scope problems solved by a detailed phone conversation with the manufacturer's service department; or mail a detailed description of the scope problem and you'll often be rewarded by a clear description of how to cure the defect.

But, whether you attempt the repair yourself or contact the manufacturer, make sure the symptoms are clear to you. Neither you nor the manufacturer's service staff can intelligently handle a vaguely described problem.

7.3 Troubleshooting Tips

Here are some initial ground rules to follow:

If the scope appears to have problems that affect both the horizontal and vertical sections, as well as faulty triggering, suspect the power supply circuits since they are common to all sections. If the vertical amplifier section exhibits problems only on certain sensitivity ranges, suspect the vertical attenuator or preamplifier; if all ranges of the vertical section appear intermittent or defective, check all stages beyond the attenuator and preamplifiers. Similarly, in the horizontal sweep and amplifier section, if the sweep is erratic on only one range, check the sweep selector switch and its associated components; if all ranges are affected, check the horizontal amplifier section. When various modes of triggering are available and triggering appears unstable, check whether triggering instability exists in all modes or just one; this input will help localize the potentially defective stage.

Remember to use all the resources available when you troubleshoot the scope. If you must remove the scope from its case to check components and voltages, remember to refer to the instruction manual for removal instructions. You may encounter dozens of screws confronting you as you start; some of the screws may hold the power transformer or CRT brackets and should not be loosened or removed. Once the chassis is exposed for inspection, use your eyes to spot burnt resistors or shorted capacitors that may appear charred. If the scope uses tubes, see that all tube filaments are lit; an open filament may be the problem. Use your nose; many overheated resistors or transformers will retain a burnt odor quite a while after power has been turned off. If you are troubleshooting an intermittent problem and power is on, use your ears to detect any slight arcing or corona from the high-voltage section. And, above all, use your brain; don't haphazardly replace parts or turn internal adjustments if you don't know what you are doing.

7.4 Replacing a Defective Component

There are several helpful rules to follow when replacing a defective component. First, try to purchase a replacement of similar physical size, working voltage, and tolerance. Before removing the defective component, sketch how the component was installed and how its leads were routed; then, when the new part is installed, try to duplicate the installation and lead dress.

Use a low-wattage (15 or 20 watt) soldering iron when working on printed-circuit boards; a larger-wattage iron will overheat the board and ruin the entire assembly. Make sure the tip of the iron is clean and properly tinned; use only enough heating time to make a good, solid solder joint. When the soldering job is finished, make sure to clip off excess wire from components to avoid shorting when the unit is replaced in its case. To avoid damage, use needle-nose pliers to act as a heat sink when soldering semiconductors (diodes or transistors).

7.5 Calibration Hints

Routine measurements with a scope involve peak-to-peak voltage readings and frequency or time values. These readings are taken from the scope with the technician's assumption that the instrument's initial factory calibrations have not changed or drifted. Unfortunately, this assumption is not valid.

Under normal temperature and humidity conditions, it is good practice to recheck scope calibration every six months for a solid-state model and every four months for a unit using tubes. Instruments operating in a hot and humid climate or experiencing mechanical abuse due to frequent shipment should be checked more frequently.

Most manufacturers offer convenient test and calibration centers where routine calibration and checkup is available at a moderate fee. Also, manufacturers' instruction manuals detail, in clear step-by-step procedure, how to calibrate the particular model scope. When you remove the scope from its case to perform internal calibration adjustments, remember to exercise extreme caution since dangerous high voltages will be exposed.

7.6 Troubleshooting Chart

The troubleshooting chart (Table 7-1) is a basic guide to fault location and repair of a scope. As previously noted, today's scopes contain many complex and novel solid-state circuits. This is not to imply that servicing is beyond the ability of the serious technician or hobbyist; rather, it strongly reinforces the point that a sound foundation of the scope's

Servicing the Oscilloscope

Table 7.1. Troubleshooting Chart

Symptom	Possible Cause	Remedy
No power, pilot light off.	On/off switch in OFF position.	Turn ON.
	Defective on/off switch.	Replace.
	AC outlet socket defective.	Try another socket.
	Open fuse.	Replace.
	Defective pilot light.	Replace.
Blows fuse when ac is applied.	Shorted power transformer.	Replace.
	Defective rectifiers; filter capacitors.	Replace.
	Short in PC board.	Visually check and remove short.
	Short in any circuit fed by power supply.	Use ohmmeter to isolate stage.
Power on but no spot or trace on CRT screen.	V or H centering controls misadjusted.	Set properly.
	Intensity control set too low.	Readjust.
	No high voltage on CRT.	Check/replace HV rectifier. Check HV filter.
	Defect in V or H amplifier stages.	Check voltages applied to CRT, deflection plates and compare with service manual.
	Defective CRT.	Replace.
Focus control does not sharpen beam.	Focus or astigmatism control open.	Replace.
	Defective CRT.	Replace.
H or V positioning control has no effect.	Defective control.	Replace.
	Defect in V or H amplifier.	Check voltages on CRT deflection plates
Sweep inoperative, no horizontal trace on any range.	Incorrect control settings.	Refer to instruction manual.
	Sweep range switch defective.	Replace.
	Open or broken leads going to sweep range switch.	Visually check and repair.
	Sweep vernier potentiometer open.	Replace.
	Sweep tube or transistor defective.	Check and replace.
Sweep OK on some ranges, but not all.	Sweep range switch defective.	Replace.
	Components on particular switch position defective or making poor contact.	Check and resolder or replace.

Table 7.1. Troubleshooting Chart (*con't.*)

Insufficient sweep.	Weak horizontal amplifier tube or transistor.	Check and replace.
	Defect in horizontal amplifier stage.	Check voltages and replace defective component.
Poor vertical deflection sensitivity, low gain.	Weak vertical amplifier tube or transistor.	Check and replace.
	Defect in vertical amplifier stages.	Check voltages, isolate defective component and replace.
	Defect in power supply.	Check power supply voltages.
	Defective probe.	Replace.
Erratic triggering.	Defect in triggering circuit.	Check components in trigger pickoff circuit and trigger circuits.

operating controls, circuitry, and calibration be well understood before the unit is removed from its case.

Caution must be exercised to avoid contact with the high-voltage power supply required for CRT operation. Use the troubleshooting chart as the start of problem analysis. Then refer to the manufacturer's instruction and/or service manual where pertinent resistance and voltage readings are shown in tables or on the schematic diagram.

7.7 Internal Adjustments

Solid-state devices, including transistors, diodes, and integrated circuits (ICs), are almost exclusively used in modern scopes; except for the CRT and possibly the high-voltage rectifier, vacuum tubes are a rarity.

Thus, the departure from vacuum tubes operated with high-temperature filaments and relatively high plate and screen voltages has offered a very positive bonus. Heat has been reduced considerably within the oscilloscope cabinet.

Technicians servicing scopes and other electronic equipment loaded with vacuum tubes are familiar with the long-range effects of heat. Dried-out electrolytic capacitors, resistors that have changed value, and frequent internal control readjustments due to component drift are not unusual.

Since many scopes still in operation employ vacuum tubes and are adversely affected by continuous exposure to heat, internal control adjustments may have to be checked and repeated. Among the internal

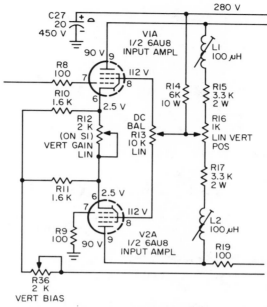

Fig. 7-1. Adjusting dc balance control.

adjustments involved are dc balance, voltage calibration, amplifier response, input attenuators, and sweep frequency.

Many oscilloscopes employ dc-coupled vertical amplifiers to permit use at very low frequencies without distortion effects. It may be necessary, at times, to adjust the dc balance control if rotating the gain control causes the trace to move up or down without an input signal. Some scopes include a coarse centering control, located within the case, to permit the beam to be located in the center of the screen (vertically or horizontally) when the appropriate control is in the center of its range. Thus, for example, the external horizontal centering control would be set at its center position; if the beam was a bit off to the left or right, the internal horizontal centering control would be adjusted to center the beam.

For example, to adjust the dc balance for the EICO model 435 wideband scope, the VERTICAL GAIN control (Fig. 7-1) is set for minimum gain. The VERT. POSITION control is set to place the horizontal trace exactly on the horizontal center line of the graticule. Now the VERTICAL GAIN control is set for maximum gain; if the trace shifts, adjust the DC BAL. control to recenter the trace. The process of varying gain and dc balance is continued until no vertical shift can be detected as vertical gain is varied over its entire range.

Before this adjustment is done, it is suggested that the scope operate for at least one-half hour until its internal case temperature is constant.

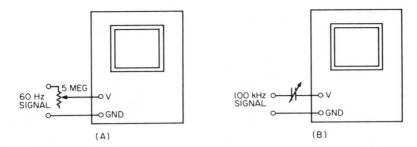

Fig. 7-2. Checking scope input resistance (A) and capacitance (B).

7.8 Checking Input Resistance and Capacitance

To determine the input resistance of a scope, a 5-Meg potentiometer is placed in series with a 60-Hz input signal and the VERT. INPUT terminals (Fig. 7-2A).

The potentiometer is set for zero resistance and the scope controls are set to observe two cycles of the 60-Hz sine wave with 10 vertical divisions deflection. Next, the potentiometer is adjusted until the scope signal is 5 units high or half the original level. At this point, the input resistance of the scope is equal to the input resistance of the potentiometer. The potentiometer is carefully disconnected to avoid upsetting its adjustment, and an ohmmeter is used to measure the resistance value.

To measure the input capacitance of a scope, a trimmer capacitor is connected in series with the VERT. INPUT of the scope. A 100-kHz sinewave signal from an audio oscillator is applied to the series circuit (see Fig. 7-2B). Now the trimmer capacitor is shorted out and the scope horizontal and vertical gain controls are set to observe two cycles at 10 divisions vertical deflection. The short across the trimmer is removed and the trimmer adjusted until a vertical signal 5 divisions high is obtained. At this point, the input capacitance of the trimmer exactly equals the scope's input capacitance. Remove the trimmer capacitor and measure its value on a capacitance bridge.

7.9 Checking Input Attenuator Compensation

Most scopes include a switch-controlled input attenuator to reduce the input signal level before application to the vertical amplifier section. As each attenuator setting is changed, the voltage divider network is varied. The stray and wiring capacitance also changes, and individual trimmers are included for each switch position to offer provisions for frequency compensation.

As shown in Fig. 7-3 for the EICO model 435, the trimmers for the 10, 100, and 1,000 positions of the VERT. ATTENUATOR are C6, C4, and C2. To adjust each trimmer, the sawtooth output available from this

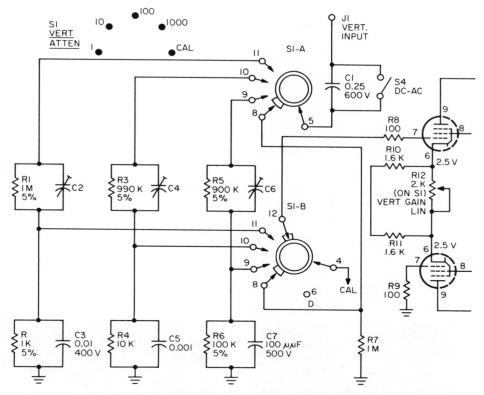

Fig. 7-3. Adjusting input attenuator compensation.

particular scope is used; other scopes use square-wave signals available from the calibration section. The SWEEP SELECTOR is set to the 1K-10K position with the SWEEP VERNIER at zero and the HORIZONTAL SELECTOR at "+ SYNC." The input is set to ac and the VERT. GAIN is set to minimum.

The attenuator is set to the 10 position, and a jumper is connected from the SAWTOOTH front-panel jack to the VERT. INPUT post. The HORIZ. GAIN control is set for a 4-inch-wide sweep. If frequency compensation is necessary, the trace on the screen will not be a sawtooth; instead it will appear with distortion at either end. Adjust trimmer C6 until a linear sawtooth line appears. A similar procedure is used to adjust the 100 and 1,000 attenuator switch settings.

7.10 Using a Square Wave to Check Compensation

A square-wave signal is convenient for checking and adjusting compensated input attenuators and the low-frequency probe of an oscilloscope. If a 100-kHz square-wave signal is applied to a vertical

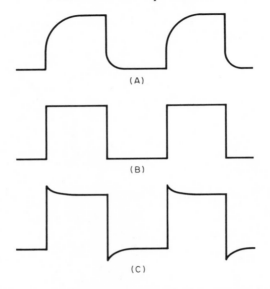

Fig. 7-4. Square-wave checking for input attenuator.

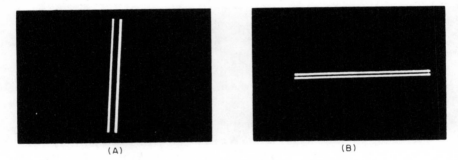

Fig. 7-5. Checking amplifier crosstalk.

Fig. 7-6. Checking amplifier phase shift.

amplifier and the scope waveform is not distorted (leading and falling edges are sharp and top and bottom are flat), it can be assumed that the amplifier is flat from 1/10 to 10 times the square-wave frequency of 100 kHz; or response is flat from 10 kHz to 1 MHz.

To check attenuator compensation, a 10-kHz square-wave signal is applied to the attenuator input. The trimmer for each particular attenuator setting is adjusted until the best square-wave pattern is observed. If trimmer capacitance is too low, the leading edge will be rounded off, as shown in Fig. 7-4A; if the trimmer value is too large, excessive peaking will appear, as shown in Fig. 7-4C. Optimum response is shown in Fig. 7-4B.

7.11 Checking the Low-Capacitance Probe

The low-capacitance probe can be checked and adjusted in the same manner as the compensated input attenuator. A square-wave signal of 10 kHz is first applied directly to the scope VERT input. A voltage reading is noted by using the voltage calibrator. Then the square-wave generator is applied to the probe and the probe output is fed to the scope input. Again, a voltage reading is noted.

7.12 Checking Crosstalk between V and H Amplifiers

When an input waveform is analyzed, it is assumed that the CRT truly displays the input signal. However, it is possible that crosstalk, or signal feedthrough, may allow some of the horizontal sweep signal to enter the vertical amplifier section and thus distort the displayed waveform.

To check crosstalk in the vertical amplifier, a square-wave signal is applied to the VERT. INPUT terminals and the HORIZ. GAIN control is set to zero. If no crosstalk exists, as the VERT. GAIN is increased, a single vertical line will be seen; if crosstalk exists, two vertical lines will be observed (see Fig. 7-5A).

To check crosstalk in the horizontal section, the square-wave signal is applied to the HORIZ. INPUT, the VERT. GAIN control is set to zero, and the HORIZ. GAIN is increased. If no crosstalk exists, a single horizontal trace will be seen; crosstalk will cause two horizontal lines to be displayed (see Fig. 7-5B).

7.13 Checking Phase Shift between V and H Amplifiers

Apply a 1-kHz signal from an audio oscillator to the VERT. INPUT and HORIZ. INPUT terminals of the scope. Adjust the gain settings for a diagonal line on the CRT. Slowly vary the frequency of the oscillator until the single diagonal line (A) begins to appear as an elipse (B); see Fig. 7-6. This frequency represents the point at which a noticeable phase shift begins to exist.

7.14 Adjusting Voltage Calibration

In the EICO 435 scope, a zener-diode controlled calibration source (Fig. 7-7) provides a precise 200-mV p-p square-wave calibration signal at line frequency. Calibration voltage adjustment is as follows:

The VERT. ATTEN. switch is set to the 10 position and the AC-DC switch to DC. A jumper is placed between the VERT. INPUT and G binding posts, and the trace is centered on the horizontal centerline of the graticule. The HORIZ. GAIN control is set to zero so the trace is merely a spot at the center of the screen. The jumper is removed and a fresh 1.5-V battery is connected between VERT. INPUT and G. The spot will be deflected away from center; adjust VERT. GAIN control for 3 cm or 3 major divisions on the vertical axis.

Next, the VERT. GAIN is not touched and the VERT. ATTEN. is switched to CAL. position. Rheostat R77 is adjusted until the square-wave calibrating voltage (appearing as a single vertical line since HORIZ. GAIN is zero) deflection is 4 cm or 4 divisions.

Now, the calibration square-wave signal is set to 200 mV, and its appearance can be viewed by turning up the HORIZ. GAIN control. Since the 200-mV signal causes a 4-cm vertical deflection, the basic vertical sensitivity setting is 50 mV/cm.

7.15 Checking Sweep Linearity

A quick check of horizontal sweep linearity can be made by applying a sine-wave signal to the VERT. INPUT terminals and then adjusting the scope controls to display four or more cycles on the CRT.

If the sweep is linear, the sine-wave cycle on the left side of the screen will be similar to those in the center and on the right (see Fig. 7-8A).

If nonlinearity exists, there will be a stretching on one side and a cramping or squeezing of the waveform on the other side (see Fig. 7-8B).

7.16 Checking V and H Sweep Settings

For convenient servicing of TV receivers, some scopes include a V and H sweep position setting, which provide 30-Hz and 7,875-Hz sawtooth waveforms, respectively. Thus, if a technician wishes to observe the vertical field, he merely switches to the V setting of the sweep selector switch and does not have to fiddle with any coarse or fine sweep adjustments.

To calibrate these two settings, the sweep selector switch is set to H and an audio oscillator is applied to the VERT. INPUT. The oscillator is set to 7,875 Hz, sync amplitude is set to zero, and the sync selector switch is set to internal. The H sweep calibration control is adjusted until the pattern on the screen displays one complete cycle.

The V setting is adjusted in a similar manner using a 30-Hz output from the audio oscillator.

Servicing the Oscilloscope

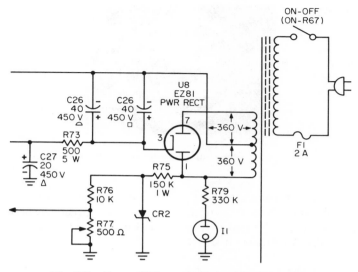

Fig. 7-7. Zener-diode controlled calibration source.

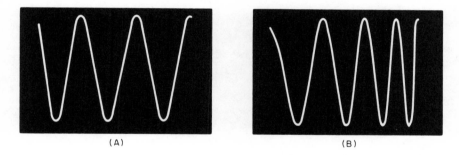

Fig. 7-8. Checking sweep linearity.

7.17 Checking Astigmatism Adjustment

Astigmatism is a defect in focusing which is evidenced on a scope by uneven focus at different sections of the scope trace. For example, a display of three sine waves on the screen may show good focus at one side of the screen but defocusing on the left; or the top areas might be sharp but the lower sections blurred.

After the FOCUS adjustment (front-panel control) is made for the sharpest display, the ASTIGMATISM control (located within the scope cabinet) is adjusted for the best compromise in sharpness.

A more efficient method to adjust ASTIGMATISM is to apply a 60-Hz sine wave directly to the VERT. INPUT terminals and also to the HORIZ. INPUT through a trimmer capacitor.

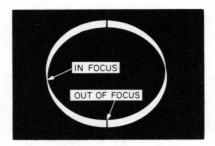

Fig. 7-9. Checking astigmatism adjustment.

Adjust the trimmer and scope gain control settings until a circular pattern is obtained. Now it is easier to observe astigmatism by checking the focus at various portions of the circle (see Fig. 7-9). Carefully adjust the ASTIGMATISM control for best overall focus.

8
selecting the oscilloscope

When shopping for a scope, the first rule of thumb is "know what you need... now or in the immediate future." Don't buy a scope loaded with features that are highly appropriate to a research lab if your intended applications will be limited to audio or TV servicing. On the other hand, it's not a good investment to purchase a low-cost, limited-bandwidth scope if you anticipate extending your TV repair shop to handle personal computers as they become more popular in the near future. Remember the old and true cliché—"time is money"; the more appropriate the scope to

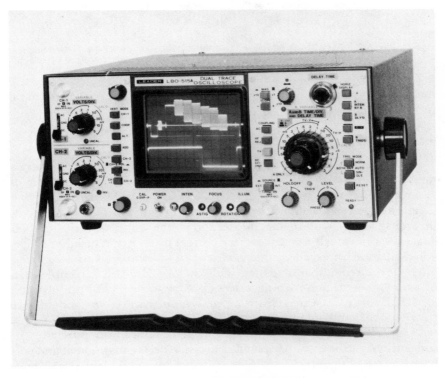

Fig. 8-1. Conventional dual-trace scope with 25-MHz bandwidth and delayed sweep. *(Courtesy* Leader Instruments Corp.)

159

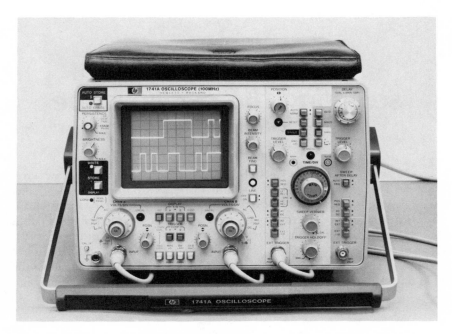

Fig. 8-2. A 100-MHz scope with variable persistence storage. *(Courtesy Hewlett-Packard.)*

the uses you have in mind, the faster your job will go and the more successful you will be.

8.1 How Much Can You Afford?

Decide how much you can afford to spend on a scope, establishing a range within which dollars you can allocate. There are a number of electronic directories available (Electronic Engineers' Master [EEM] and Electronic Design's Gold Book) that list manufacturers of test equipment, including scopes, with their addresses and phone numbers. Write for their data sheets and brochures, and request current price lists. When they arrive, you will have an excellent reference source to compare various models offered by different manufacturers with key specs detailed. A complete list of scope products and manufacturers appears at the end of this chapter.

Early in your search, make a decision on whether you should try a conventional self-contained scope (Figs. 8-1 and 8-2) or a mainframe type with optional plug-ins (Figs. 8-3 and 8-4). Once you purchase the conventional scope, you have made a decision on the instrument and will have to live with the specs offered. This is quite all right if you have carefully analyzed your needs. On the other hand, if you purchase a mainframe and one particular time base, for example, you can upgrade or

Fig. 8-3. Mainframe with four plug-in units including two 200-MHz vertical amplifiers and variable-delay time bases. *(Courtesy* Tektronix.)

change specs by simply purchasing a different time base plug-in. The flexibility of a plug-in scope over a conventional scope is offset by the additional cost of the mainframe and separate plug-ins versus the cost of a single instrument.

8.2 Conventional Scopes versus Plug-in

What are some guidelines to help decide whether a mainframe and plug-ins are preferable? Assume a research lab will include six technicians involved in a variety of applications, ranging from high-speed logic to biomedical equipment. Rather than purchase quite a few conventional scopes with different specs for varied needs, only six mainframe scopes would be necessary with a number of plug-ins. As different projects are undertaken, the various plug-ins, for low-frequency amplifiers or high-speed time base, could be selected and inserted. On the other hand, a service lab dedicated to TV and stereo repair might be unwise to invest in mainframes since only a few plug-in variations would even be considered; here a wideband conventional scope for each service bench would be a better choice. If a rack-mounted model is required, they are also available (see Fig. 8-5).

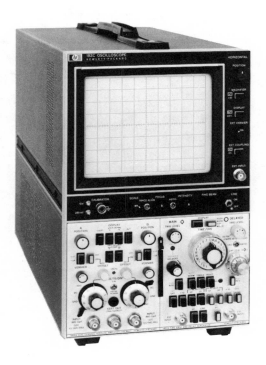

Fig. 8-4. Mainframe with two 100-MHz vertical amplifiers and delayed time base. *(Courtesy* Hewlett-Packard.)

8.3 Factors Affecting Buying Decision

What features are important in the buying decision? Among the key features are (1) CRT; (2) sweep—recurrent or triggered? (3) time base—is a calibrated time base important? (4) bandwidth—how much is enough? (5) sensitivity—how much is really necessary? (6) portability—will the scope be used in the field where lightweight and battery-powered features are absolutely essential? and (7) key specs—what do they really mean?

8.4 The CRT

The cathode-ray tube, or CRT, is what you will be looking at. Most triggered scopes use a P31 phosphor since it offers medium persistence, desirable for most applications; an aluminized version of the P31, the P31A, provides more brightness and contrast in its green trace. Some scope tubes use a P11 phosphor which displays a blue trace.

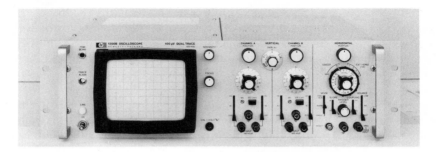

Fig. 8-5. Dual-trace scope convenient for rack mounting. *(Courtesy Hewlett-Packard.)*

Although almost all scope front panels indicate a rectangular CRT format, it is not uncommon to find round CRTs used with a rectangular mask. The most popular size rectangular scopes display an 8 × 10 cm area; the vertical or X axis is 8 cm high while the horizontal length is 10 cm. The graticule grid, placed in front of the screen to facilitate peak-to-peak readings, is generally spaced in 1-cm divisions.

A sharp trace is necessary in a wideband scope since applications generally include investigation of fast pulses. Many scopes include an astigmatism control, often adjustable with a small screwdriver rather than with a front panel knob, which is set to display a tiny pinpoint for the beam instead of an elongated dot.

8.5 Scope Bandwidth

Not all scope manufacturers express bandwidth in a standard manner. Generally, the vertical amplifier frequency response (indicated as the scope bandwidth) is given from dc or some low frequency to its higher-frequency limit. The standard tolerance for the dropoff in vertical gain is 3 dB; +3 dB indicates as much as 1.4 times the normal or reference level, while −3 dB indicates a voltage level 0.7 as great as the standard level. Thus, over a wide frequency range, a scope's vertical amplifier could deliver 1-V output at midfrequency, 1.4 V somewhere above midfrequency, and 0.7 V near the low-frequency region, and could maintain a +3 dB or −3 dB variation over the frequency range.

To achieve wide bandwidth, scope designers either eliminate coupling capacitors or else use large values to allow low frequencies to pass with negligible attenuation. To minimize attenuation at high frequencies due to stray and circuit capacitances, "peaking" circuits are sometimes used. This arrangement extends the high-frequency response of a scope but may cause a serious peak followed by a sharp dip at the peaking frequency; under these conditions, a manufacturer can claim bandwidth extended to, say, 15 MHz but the response would not be flat or within 3 dB over the entire frequency band. A different approach to reduce attenuation of high-frequency signals involves the use of multiple, low-gain vertical amplifier stages. Generally, with this design, the scope manufacturer can boast, with integrity, that the response is within 3 dB, without peaks or dips, to the higher frequencies, 15 MHz as an example.

Thus, two manufacturers can claim "response to 15 MHz" in their specs. The first scope could indeed display signals up to 15 MHz, but signals at 10 MHz might be considerably higher in amplitude due to excessive peaking. The second scope, not using peaking, would also display the 15-MHz signals, but 10-MHz signals would experience the same gain due to the flat response of the vertical amplifier stage. One point to note: peaking is an acceptable method of extending vertical amplifier response used by many scope manufacturers. The overall frequency response can be quite flat, within 3 dB or less, and will be so stated in the manufacturers' specs. As long as the specs indicate "within 3 dB" or "within 1 dB," performance is defined. Be wary of vague specs that boast "displays signals up to 15 MHz."

Now, how wide a bandwidth do you need? If you are a service technician concentrating on audio and TV work, a 5-MHz scope would be sufficient to display all waveforms you will encounter, including the 3.58-MHz color burst signals. Suppose, however, that you also intend to service CB equipment. It's doubtful that you'll see any 27-MHz RF signal on the 5-MHz scope; the response of the vertical amplifier section is just too low. So you might find it necessary to shop for a 30- or 50-MHz scope; yes, it's more expensive but necessary if you wish to speed your CB troubleshooting and make a profit.

If you are really serious about tracing pulses in microcomputers or the rapidly growing field of personal computers, then a wide bandwidth scope is a definite "must" (see Fig. 8-6). The short-duration pulses involved in such equipment demand vertical amplifiers with very short risetime (risetime is the period required for the leading edge of a pulse to rise from 10 to 90 percent of its final magnitude). A simple rule of thumb for relating risetime to frequency scope bandwidth is: divide 0.35 by the risetime (in seconds) and the result will be the 3-dB bandwidth in hertz. For example, if the computer involves pulses with 10-nsec (10^{-9} sec) risetime, what bandwidth scope would be required? Divide 0.35 by 10^{-9} and the answer would be 35,000,000 or a 35-MHz scope.

8.6 Sweep Range

Once a decision has been made on the bandwidth requirement of the scope to be purchased, proceed to other relatively important parameters. Sweep range is often overlooked, yet this can be an annoying omission in decision making after the purchase. Here's why. Assume you buy a 5-MHz bandwidth scope and set it up on your service bench to display the 3.58-

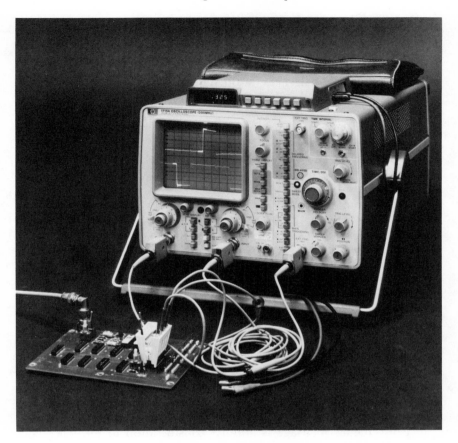

Fig. 8-6. For computer analysis and troubleshooting, a dual-trace 200-MHz scope with direct readout for time-interval measurements is most convenient. *(Courtesy* Hewlett-Packard.)

MHz color burst signal. All goes well and a vertical signal sufficient to fill the screen appears. You adjust the sweep range switch and find the highest sweep is only 100 kHz; you lock the 3.58-MHz signal and observe a long series (35) of sine waves on the screen. If consideration had been given to sweep range, with color burst signals as an essential measurement, a scope with at least a 1-MHz sweep would have been selected. It is obviously difficult, if not impossible, to observe distortion in the 3.58-MHz signal when 35 of them are shown side-by-side on a 10-cm-wide screen; only four would be displayed with a 1-MHz sweep range.

8.7 Deflection Sensitivity

Another consideration in the choice of a scope is the vertical deflection sensitivity—or how small a signal can be applied to the vertical input and produce an adequate-size display. Generally, a display that is equal to or greater than two divisions on the vertical graticule is considered useful.

Most scopes are rated, in vertical amplifier sensitivity, as the input voltage required to deflect the beam vertically one division (generally 1 cm). Thus, a scope rated at 5 mV indicates that a signal as low as 5 mV will produce a 1-cm-high display. However, do not expect to be able to see much in the way of distortion with a 1-cm-high display; remember, it is desirable to have a display as large as 2 cm or better. So if you expect to deal with signals as low, for example, as 5 mV, purchase a scope with at least 2 mV sensitivity so the display will be over two divisions in height and will allow waveform inspection.

When comparing sensitivity specs, make sure ratings are in common terms. Most sensitivity specs are indicated in rms; however, some scope manufacturers list their specs in peak-to-peak (p-p) ratings. To convert one to the other, for comparison purposes, merely multiply the rms rating by 2.8 (or use 3 for convenience) and then compare. For example, which has greater sensitivity, scope A with 5 mV rms or scope B with 10 mV p-p sensitivity figure? Scope B—it requires less than 3 mV rms to deflect the beam 1 cm vertically. However, it is also clear that a scope with 5 mV p-p sensitivity is more sensitive than another scope with 5 mV rms sensitivity—about three times more sensitive.

Also bear in mind the type of circuits you may regularly be checking. If a low-capacitance probe is to be used often, remember there is a 10:1 attenuation factor to be considered. This may influence the sensitivity requirement of the scope and point to the need for a better sensitivity instrument than you had in mind.

Convenience in usage is another consideration to influence your choice. Most modern scopes include a front-panel vertical amplifier control marked in volts/cm or volts/division. To make a peak-to-peak voltage reading, connect to the appropriate test point, sync the signal, set the vertical input control to the setting where the entire vertical signal can be seen, count the number of divisions and multiply it by the input control setting. For example, if the display is four divisions high and the input selector switch is on 1 V/division, the waveform is 4 V peak-to-peak. Scopes with such simplified calibration schemes require that the input switch be set in the calibrated or cal position. It is also necessary to consider that the voltage readings will be accurate if the signals under test are within the bandwidth of the scope's vertical amplifiers.

8.8 Time Base

Recurrent sweep scopes are almost obsolete by now, although some lower-priced models still use this sweep approach (see Fig. 8-7). Triggered scopes, with their calibrated time base, allow the user to rapidly view and measure various segments of a composite signal. The sweep control

Selecting the Oscilloscope 167

Fig. 8-7. A low-priced, 4.5-MHz recurrent sweep, single-trace scope is adequate for many educational and servicing tasks. *(Courtesy* EICO Electronic Instruments Co., Inc.)

determines how long it takes the beam to move from the left- to right-hand side of the screen. Thus, by adjusting the time base, it is possible to condense or expand the waveform under observation. Let's take the composite TV signal, assuming it was desirable to inspect the total horizontal line, including video detail, sync pulse, color burst and the blanking pedestal. One horizontal line scan occurs every 63.6 μsec; if a 10 μsec/division time base is selected, the total horizontal signal will be shown in slightly more than six horizontal divisions. A technician can conveniently examine the waveform and check for compressed video or clipped sync. Should it be necessary to check the burst signal, the time base can be set to 2 μsec/division and the signal would effectively be expanded horizontally by a factor of five. Now it is considerably easier to view the color burst signal and check for attenuation or distortion.

Since the time base on most recurrent scopes is accurately calibrated, it is a simple matter to set the time base control to display one complete cycle and read the time required between start and finish; by taking the reciprocal or dividing this time into 1, the signal frequency can be found.

8.9 Dual-Trace Scopes

In the past few years, there has been a surge in the sales of dual-trace scopes. Reason? The rapidly growing number of computers (mini, micro and personal) that require precisely timed pulses for proper operation. Signal tracing a TV set or an audio amplifier is quite adequately handled

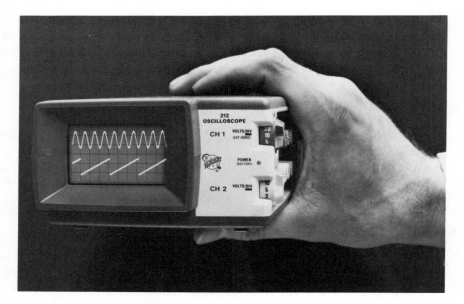

Fig. 8-8. Portable, battery-operated dual-trace scope weighs only 3½ lb. *(Courtesy* Tektronix.)

with a conventional, single-trace scope. But pulse circuits for counting, division, clock synchronization, and other precision functions are conveniently monitored when input and output signals can be viewed simultaneously. Similarly, a dual-trace scope rapidly compares signal amplitude and waveforms in stereo receivers, when both channels can be observed at the same time.

8.10 The Final Decision

When you have made all the necessary decisions regarding application, portability (see Fig. 8-8), bandwidth, sensitivity, sweep range, and the like, you are now ready to make the purchase. Visit several suppliers of electronic components and test equipment and ask questions—don't be embarrassed. Come equipped with your questions written on a sheet of paper with room to jot down answers. Try to determine early in the discussion whether the salesman is (1) competent and capable of offering objective comments on various manufacturers' products, (2) honest in his responses and not "steering" you to what he wishes to sell, and (3) informed enough not only to answer your questions but to offer additional inputs to help you in your selection. Check the guarantees offered by the manufacturer and the store; who is responsible if the scope fails a day after it is put into service or, worse yet, does not function when it is removed from the carton in your home or shop? Ask the salesman to allow you to "fiddle" with the front-panel controls of the various scopes you will select from. You may find one brand more convenient to your particular likes than the others; several scope manufacturers have invested considerable expense in human engineering design of their front panels to make the instrument easier to use.

If the decision is now down to one or two brands or models, ask the salesman to permit you to read the instruction/application manual supplied. All elements being equal, a well-prepared manual could tip the decision.

When you make the purchase, make sure all guarantees and promises are in writing. Keep the sales receipt, guarantees, and instruction manual together in a convenient place where they can be found quickly. Make sure you carefully read the instruction manual before you remove the scope from the carton and plug it into an ac outlet. A little patience, with a lot of study, will eventually provide you with the most versatile servicing tool on your bench.

A list of typical recurrent-sweep scopes appears in Table 8-1. Only a limited number of models are shown for illustration; many others are available from these and other manufacturers. Table 8-2 lists some of the many available triggered-sweep scopes on the market, both single- and dual-trace. Prices shown are approximate; make sure to consult more current price lists as you complete your buying analysis. Table 8-3 is a complete list of scope manufacturers with their mailing addresses and phone numbers to assist your search for their literature and application information (Courtesy Electronic Design's Gold Book, Hayden Publishing Co., Rochelle Park, N.J.).

8.11 Renting and Leasing Options

A relatively new approach to acquiring oscilloscopes (and other test equipment) is to rent or lease rather than purchase them.

The three alternatives include straight rental, rental with option to buy, and leasing. With straight rental, often a short-term situation involving several weeks to a few months, monthly fees are about 10 percent of purchase price. The scope is borrowed for the set fee and returned when it is no longer needed.

Rental with option to buy permits the borrower to use part of the rental fee toward the eventual purchase of the scope at the completion of the rental period. The general approach is to allow 50 percent of the rental fee to be applied toward the instrument purchase if the rental is relatively short-term, six months or so. However, if the instrument is borrowed for a longer period, and monthly fees are paid regularly, up to 75 percent of the total rental fee may be applied toward the purchase.

The third alternative, leasing, is similar to rental but extends over a longer period; two to five years typically. Monthly leasing fees are lower than rental but the maintenance is the responsibility of the customer. At the completion of the leasing term, the customer may renew his lease, buy the equipment at a reduced price, or simply return the scope.

The advantage of an outright purchase is obvious: since you or your company owns the scope, tax deductions are possible for depreciation. Depreciation is a serious consideration since a lab scope can become relatively obsolete in a few short years. For a small business, purchasing several scopes may require a substantial cash outlay and can involve heavy interest costs for bank loans. In addition, the owner of a scope is responsible for its periodic maintenance, calibration, and repair expenses.

The key factor in deciding whether to rent or buy is the inflation rate. If you purchase a scope, you pay with today's dollars, the value of which is dropping, and you also receive a tax break. When you lease, if the dollar still continues to drop, you pay a fixed monthly fee in dollars that are steadily declining in value. Therefore, it is possible to lease a scope over a several-year period, pay more dollars over this time than the original purchase price, and still be ahead because of the lower value of the dollar at the end of the lease period.

The middle alternative, renting, appears to be an optimum option over a short-term period. If a small lab or service shop needs several scopes for a special project or flood of work that will span a month or so, it is wise to rent and return the equipment when the crisis is over.

The instrument rental and leasing companies listed in Table 8-4 can supply current fees and details on this new trend.

Table 8.1. Typical Recurrent-Sweep Oscilloscopes Available

Manufacturer	Model	Bandwidth (MHz)	CRT Size (in.)	Sensitivity (mV/div.)	Sweep Range (Hz-kHz)	Approx. Price ($)
B & K Precision	1403A	5	3	10	10-110	250
Eico	427	0.5	5	10	10-100	250
	460	4.5	5	35	10-100	200
	435	4.5	3	50	10-100	250
Leader	LBO-310A	4	3	20	10-100	200
	LBO-511	10	5	20	10-100	300

Table 8.2. Typical Triggered Oscilloscopes Available

Manufacturer	Model	Bandwidth (MHz)	Dual-Trace	Sensitivity (mV/div.)	Sweep Time μsec to sec	Mag.	Probes (Included)	Approx. Price ($)
B & K Precision	1461	10	No	10	1.0–0.5	5X	Yes	500
	1472C	15	Yes	10	0.5–0.5	5X	Yes	800
Ballantine	1010A	15	Yes	2				700
Eico	480	10	No	10	0.1–0.5	–	Yes	450
Gould	OS245A	10	Yes	5	1.0–0.5	10X	Yes	600
	OS260	15	Dual-Beam	5	0.5–0.2	10X	Yes	900
Heath	IO-4555	10	No	10	0.2–0.2	5X	No	330 (kit)
	IO-4560	5	No	100	0.2–0.2	–	No	575
	IO-4510	15	Yes	1	0.1–0.2	5X	No	800
Hewlett-Packard	1740A	100	Yes	5	0.05–2.0	10X	No	2,250
	1220A	15	Yes	2	0.1–0.5		Yes	795
Hickok	515	15	Yes	10	0.5–0.2	5X	Yes	500
	532	30	Yes	10	0.05–2.0	4X	Yes	1,000
Leader	LBO-507	20	No	10	0.5–0.2	5X	Yes	550
	LBO-520	30	Yes	5	0.2–0.5	10X	Yes	1,050
Nonlinear Systems	MS215	15	Yes	10	0.1–0.5	–	No	450
Philips	PM3211	15	Yes	2	0.5–0.2	5X	Yes	975
	PM-3226	15	Yes	2	0.5–0.2	5X	Yes	875
Sencore	PS29	8	No	10	0.2–0.1	5X	Yes	
	PS163	8	Yes	5	0.1–0.1	5X	Yes	800
Simpson	452	15	Yes	5	0.2–0.5	5X	No	700
Tektronix	T921	15	No	2	0.2–0.5	10X	Yes	900
	T922	15	Yes	2	0.2–0.5	10X	Yes	
Viz	WO-527A	15	No	10	0.5–0.5	10X	Yes	525
	WO-555	15	Yes	10	0.5–0.5	10X	Yes	750

Table 8.3. Oscilloscope Manufacturers

OSCILLOSCOPE PLUG-INS

BEC Inc, Certified Calibration Labs, Dept G, 2709 N Broad St Phila PA 19132	(215)229-9800
Continental Resources Inc, 175G Middlesex Turnpike Bedford MA 01730	(617)275-0850
Dumont Oscilloscope Labs Inc, 40G Fairfield Pl W Caldwell NJ 07006	(201)575-8666
Electro Rent Corp, 4131G Vanowen Pl Burbank CA 91505	(800)232-2173
General Atronics Corp, Dept G, 1200 E Mermaid Ln Phila PA 19118	(215)248-3700
Hewlett-Packard, Corporate Div, 1501G Page Mill Rd Palo Alto CA 94304	(415)856-1501
Julie Research Labs Inc Dept G, 211 W 61 St NY NY 10023	(212)245-2727
Livingston Hire Ltd, Hire Elec Instru, 27–29G Camden Rd Shirley House, London NW1 9NR, England	01 2673262
Tektronix Inc, Dept G, PO Box 500 Beaverton OR 97077	(503)644-0161
Tracor Northern Inc, Dept G, 2551 W Beltline Hwy Middleton WI 53562	(608)831-6511
Tucker Elecs Co, Dept G, PO Box 401060 Garland TX 75040	(214)348-8800

OSCILLOSCOPES, CATHODE RAY

Aul Instrus Inc, 1400G Plaza Ave New Hyde Pk NY 11040	(516)437-2400
Ballantine Labs Inc, Dept G, PO Box 97 Boonton NJ 07005	(201)335-0900
BEC Inc, Certified Calibration Labs, Dept G, 2709 N Broad St Phila PA 19132	(215)229-9800
Dumont Oscilloscope Labs Inc, 40G Fairfield Pl W Caldwell NJ 07006	(201)575-8666
Dynascan Corp, B & K Precision, Dept G, 6460 W Cortland Chicago IL 60635	(312)889-9087
Electroplan Ltd, Dept G, PO Box 19 Orchard Rd, Royston, Herts SG8 5HH, England	Royston(0763)41171
Farnell Instrus Ltd, Dept G, Sandbeck Way, Wetherby Yorks LS22 4DH, Great Britain	0937 63541
General Atronics Corp, Dept G, 1200 E Mermaid Ln Phila PA 19118	(215)248-3700
Gould Inc, Instru Div Oscilloscope Prods, 3631G Perkins Av Cleveland OH 44114	(216)361-3316
Hameg, Dept G, 5–9 Av de la Republic, 94800 Villejuif, France	678.09.98
Heath Co, Dept G, Benton Harbor MI 49022	(616)982-3200
Hewlett-Packard, Corporate Div, 1501G Page Mill Rd Palo Alto CA 94304	(415)856-1501
Hickok Electrical Instru Co, Indl Instru, 10514G Dupont St Cleveland OH 44108	(216)541-8060
Infodex Inc, 7G Cherry Av Waterbury CT 06702	(203)757-9291
ITT Elecs & Indl Components Grp, Dept G, Av Louise 480, B-1050 Brussels, Belgium	649 96 20
Lawtronics Ltd, 139G High St, Edenbridge Kent TN8 5AX, England	0732-865191
Leader Instrus Corp, 151G Dupont St Plainview NY 11803	(516)822-9300
Lectrotech Inc, Dept G, 5810 N Western Av Chicago IL 60659	(312)769-6262
Livingston Hire Ltd, Hire Elec Instru, 27–29G Camden Rd Shirley House, London NW1 9NR, England	01 2673262
Philips Test & Measuring Instrus, 85G McKee Dr Mahwah NJ 07430	(201)529-3800
Promax, Engrg Dept, Galileo 249G, Barcelona 28, Spain	(93)330-9751
RTC La Radiotechnique-Compelec, Dept G, 130 Ave Ledru-Rollin, 75540 Paris Cedex 11, France	355-4499
Scientific-Atlanta, Spectral Dynamics, Dept G, PO Box 671 San Diego CA 92112	(714)268-7100
Scopex Instrus Ltd, Dept G, Pixmore Indl Est, Letchworth Herts, England	04626-72771
Simpson Electric Co, Katy Inds, 853G Dundee Av Elgin IL 60120	(312)697-2260
Snap On Tools Corp, 2801G 80 St Kenosha WI 53140	(414)654-8681

Table 8.3. Oscilloscope Manufacturers (con't.)

Soltec Corp, 11684G Pendleton St Sun Valley CA 91352	(213)767-0044
Techni-Tool Inc, 5G Apollo Rd Plymouth Meeting PA 19462	(215)825-4990
Tektronix Inc, Dept G, PO Box 500 Beaverton OR 97077	(503)644-0161
Thomson-CSF, Dept G, 23 rue de Courcelles-BP 96-08, 75362 Paris Cedex 08, France	(1)25652-52
Tucker Elecs Co, Dept G, PO Box 401060 Garland TX 75040	(214)348-8800
VIZ Mfg Co, Dept G, 335 E Price St Phila PA 19144	(215)844-2626
Vu-Data Corp, 7170G Convoy Court San Diego CA 92111	(714)279-6572
Wavetek Indiana Inc, Dept G, 66 N 1 Av Beech Grove IN 46107	(317)783-3221

OSCILLOSCOPES, MULTIPLE CHANNEL

Ballantine Labs Inc, Dept G, PO Box 97 Boonton NJ 07005	(201)335-0900
Continental Resources Inc, 175G Middlesex Turnpike Bedford MA 01730	(617)275-0850
Dumont Oscilloscope Labs Inc, 40G Fairfield Pl W Caldwell NJ 07006	(201)575-8666
Electro Rent Corp, 4131G Vanowen Pl Burbank CA 91505	(800)232-2173
General Atronics Corp, Dept G, 1200 E Mermaid Ln Phila PA 19118	(215)248-3700
Gould Inc, Instru Div Oscilloscope Prods, 3631G Perkins Av Cleveland OH 44114	(216)361-3316
Heath Co, Dept G, Benton Harbor MI 49022	(616)982-3200
Hewlett-Packard, Corporate Div, 1501G Page Mill Rd Palo Alto CA 94304	(415)856-1501
Infodex Inc, 7G Cherry Av Waterbury CT 06702	(203)757-9291
ITT Elecs & Indl Components Grp, Dept G, Av Louise 480, B-1050 Brussels, Belgium	649 96 20
Livingston Hire Ltd, Hire Elec Instru, 27-29G Camden Rd Shirley House, London NW1 9NR, England	01 2673262
Nicolet Instru Corp, 5225G Verona Rd Madison WI 53711	(608)271-3333
Norland Corp, Dept G, Rte. 4 Norland Dr Ft Atkinson WI 53538	(414)563-8456
Philips Test & Measuring Instrus, 85G McKee Dr Mahwah NJ 07430	(201)529-3800
Simpson Electric Co, Katy Inds, 853G Dundee Av Elgin IL 60120	(312)697-2260
Tektronix Inc, Dept G, PO Box 500 Beaverton OR 97077	(503)644-0161
Tele-Measurements Inc, 145G Main Av Clifton NJ 07014	(201)473-8822
Tucker Elecs Co, Dept G, PO Box 401060 Garland TX 75040	(214)348-8800

OSCILLOSCOPES, MULTI-TRACE

Ballantine Labs Inc, Dept G, PO Box 97 Boonton NJ 07005	(201)335-0900
BEC Inc, Certified Calibration Labs, Dept G, 2709 N Broad St Phila PA 19132	(215)229-9800
Continental Resources Inc, 175G Middlesex Turnpike Bedford MA 01730	(617)275-0850
Dumont Oscilloscope Labs Inc, 40G Fairfield Pl W Caldwell NJ 07006	(201)575-8666
Dynascan Corp, B & K Precision, Dept G, 6460 W Cortland Chicago IL 60635	(312)889-9087
Electroplan Ltd, Dept G, PO Box 19 Orchard Rd, Royston, Herts SG8 5HH, England	Royston(0763)41171
Electro Rent Corp, 4131G Vanowen Pl Burbank CA 91505	(800)232-2173
General Atronics Corp, Dept G, 1200 E Mermaid Ln Phila PA 19118	(215)248-3700
General Electric Co, Quick Rental Instrus, 1G River Rd-Bldg 6-328 Schenectady NY 12345	(518)385-9950
Gould Advance Ltd, Instrus, Dept G, Roebuck Rd, Hainault, Essex 1G6 3UE, England	01-500-1000
Gould Inc, Biomation, 4600G Old Ironsides Dr Santa Clara CA 95050	(408)988-6800

Selecting the Oscilloscope

Table 8.3. Oscilloscope Manufacturers (con't.)

Gould Inc, Instru Div Oscilloscope Prods, 3631G Perkins Av Cleveland OH 44114	(216)361-3316
Hameg, Dept G, 5-9 Av de la Republic, 94800 Villejuif, France	678.09.98
Heath Co, Dept G, Benton Harbor MI 49022	(616)982-3200
Hewlett-Packard, Corporate Div, 1501G Page Mill Rd Palo Alto CA 94304	(415)856-1501
Hickok Electrical Instru Co, Indl Instru, 10514G Dupont St Cleveland OH 44108	(216)541-8060
Infodex Inc, 7G Cherry Av Waterbury CT 06702	(203)757-9291
ITT Elecs & Indl Components Grp, Dept G, Av Louise 480, B-1050 Brussels, Belgium	649 96 20
Lawtronics Ltd, 139G High St, Edenbridge Kent TN8 5AX, England	0732-865191
Leader Instrus Corp, 151G Dupont St Plainview NY 11803	(516)822-9300
Lectrotech Inc, Dept G, 5810 N Western Av Chicago IL 60659	(312)769-6262
Livingston Hire Ltd, Hire Elec Instru, 27-29G Camden Rd Shirley House, London NW1 9NR, England	01 2673262
Logotron AG, 8805 Richterswil Switzerland	
Philips Test & Measuring Instrus, 85G McKee Dr Mahwah NJ 07430	(201)529-3800
Promax, Engrg Dept, Galileo 249G, Barcelona 28, Spain	(93)330-9751
Racal Grp Svcs Ltd, Racal Instrus Ltd, 21G Market Pl, Wokingham, Berkshire RG11 1AJ, England	69811
Rental Elecs Inc, 2445G Farber Pl Palo Alto CA 94303	(415)324-8080
Scopex Instrus Ltd, Dept G, Pixmore Indl Est, Letchworth Herts, England	04626-72771
Sencore Inc, 3200G Sencore Dr Sioux Falls SD 57107	(605)339-0100
Soltec Corp, 11684G Pendleton St Sun Valley CA 91352	(213)767-0044
SPI-ITT, Metrix, Dept G, Chemin de la Croix Rouge PO Box 30, 74010 Annecy Cedex, France	(50)52-8102
Tektronix Inc, Dept G, PO Box 500 Beaverton OR 97077	(503)644-0161
Tele-Measurements Inc, 145G Main Av Clifton NJ 07014	(201)473-8822
Texscan Corp, Dept G, 2446 N Shadeland Av Indianapolis IN 46219	(317)357-8781
Tucker Elecs Co, Dept G, PO Box 401060 Garland TX 75040	(214)348-8800
Vu-Data Corp, 7170G Convoy Court San Diego CA 92111	(714)279-6572
Wavetek Indiana Inc, Dept G, 66 N 1 Av Beech Grove IN 46107	(317)783-3221

OSCILLOSCOPES, PORTABLE

Ballantine Labs Inc, Dept G, PO Box 97 Boonton NJ 07005	(201)335-0900
Bruel & Kjaer, Naerum Hovedgade 18G, DK-2850 Naerum, Denmark	02-800500
Continental Resources Inc, 175G Middlesex Turnpike Bedford MA 01730	(617)275-0850
Dumont Oscilloscope Labs Inc, 40G Fairfield Pl W Caldwell NJ 07006	(201)575-8666
Dynascan Corp, B & K Precision, Dept G, 6460 W Cortland Chicago IL 60635	(312)889-9087
Electroplan Ltd, Dept G, PO Box 19 Orchard Rd, Royston, Herts SG8 5HH, England	Royston(0763)41171
Electro Rent Corp, 4131G Vanowen Pl Burbank CA 91505	(800)232-2173
General Atronics Corp, Dept G, 1200 E Mermaid Ln Phila PA 19118	(215)248-3700
General Electric Co, Quick Rental Instrus, 1G River Rd-Bldg 6-328 Schenectady NY 12345	(518)385-9950
Gould Advance Ltd, Instrus, Dept G, Roebuck Rd, Hainault, Essex 1G6 3UE, England	01-500-1000
Gould Inc, Instru Div Oscilloscope Prods, 3631G Perkins Av Cleveland OH 44114	(216)361-3316
Hewlett-Packard, Corporate Div, 1501G Page Mill Rd Palo Alto CA 94304	(415)856-1501

Table 8.3. Oscilloscope Manufacturers (con't.)

Hickok Electrical Instru Co, Indl Instru, 10514G Dupont St Cleveland OH 44108	(216)541-8060
Infodex Inc, 7G Cherry Av Waterbury CT 06702	(203)757-9291
ITT Elecs & Indl Components Grp, Dept G, Av Louise 480, B-1050 Brussels, Belgium	649 96 20
Lawtronics Ltd, 139G High St, Edenbridge Kent TN8 5AX, England	0732-865191
Lectrotech Inc, Dept G, 5810 N Western Av Chicago IL 60659	(312)769-6262
Livingston Hire Ltd, Hire Elec Instru, 27-29G Camden Rd Shirley House, London NW1 9NR, England	01 2673262
Non-Linear Sys Inc, 533G Stevens Av Solana Beach CA 92075	(714)755-1134
Philips Test & Measuring Instrus, 85G McKee Dr Mahwah NJ 07430	(201)529-3800
Promax, Engrg Dept, Galileo 249G, Barcelona 28, Spain	(93)330-9751
Racal Grp Svcs Ltd, Racal Instrus Ltd, 21G Market Pl, Wokingham, Berkshire RG11 1AJ, England	69811
Rental Elecs Inc, 2445G Farber Pl Palo Alto CA 94303	(415)324-8080
Scopex Instrus Ltd, Dept G, Pixmore Indl Est, Letchworth Herts, England	04626-72771
Sencore Inc, 3200G Sencore Dr Sioux Falls SD 57107	(605)339-0100
Simpson Electric Co, Katy Inds, 853G Dundee Av Elgin IL 60120	(312)697-2260
Sinclair Radionics Inc, Dept G, 66G Mt Prospect Av Clifton NJ 07015	(201)472-7600
SPI-ITT, Metrix, Dept G, Chemin de la Croix Rouge PO Box 30, 74010 Annecy Cedex, France	(50)52-8102
Techni-Tool Inc, 5G Apollo Rd Plymouth Meeting PA 19462	(215)825-4990
Tektronix Inc, Dept G, PO Box 500 Beaverton OR 97077	(503)644-0161
Tucker Elecs Co, Dept G, PO Box 401060 Garland TX 75040	(214)348-8800
VIZ Mfg Co, Dept G, 335 E Price St Phila PA 19144	(215)844-2626
Vu-Data Corp, 7170G Convoy Court San Diego CA 92111	(714)279-6572

OSCILLOSCOPES, PORTABLE, BATTERY OPERATED

Ballantine Labs Inc, Dept G, PO Box 97 Boonton NJ 07005	(201)335-0900
Bruel & Kjaer, Naerum Hovedgade, 18G, DK-285C Naerum, Denmark	02-800500
Continental Resources Inc, 175G Middlesex Turnpike Bedford MA 01730	(617)275-0850
Dumont Oscilloscope Labs Inc, 40G Fairfield Pl W Caldwell NJ 07006	(201)575-8666
Electroplan Ltd, Dept G, PO Box 19 Orchard Rd, Royston, Herts SG8 5HH, England	Royston(0763)41171
General Atronics Corp, Dept G, 1200 E Mermaid Ln Phila PA 19118	(215)248-3700
General Electric Co, Quick Rental Instrus, 1G River Rd-Bldg 6-328 Schenectady NY 12345	(518)385-9950
Hewlett-Packard, Corporate Div, 1501G Page Mill Rd Palo Alto CA 94304	(415)856-1501
Infodex Inc, 7G Cherry Av Waterbury CT 06702	(203)757-9291
ITT Elecs & Indl Components Grp, Dept G, Av Louise 480, B-1050 Brussels, Belgium	649 96 20
Lawtronics Ltd, 139G High St, Edenbridge Kent TN8 5AX, England	0732-865191
Lectronic Research Labs Inc, 1423G Ferry Av Camden NJ 08104	(609)541-4200
Livingston Hire Ltd, Hire Elec Instru, 27-29G Camden Rd Shirley House, London NW1 9NR, England	01 2673262
Non-Linear Sys Inc, 533G Stevens Av Solana Beach CA 92075	(714)755-1134
Philips Test & Measuring Instrus, 85G McKee Dr Mahwah NJ 07430	(201)529-3800
Rental Elecs Inc, 2445G Farber Pl Palo Alto CA 94303	(415)324-8080
Scopex Instrus Ltd, Dept G, Pixmore Indl Est, Letchworth Herts, England	04626-72771
Sinclair Radionics Inc, Dept G, 66G Mt Prospect Av Clifton NJ 07015	(201)472-7600
SPI-ITT, Metrix, Dept G, Chemin de la Croix Rouge PO Box 30, 74010 Annecy Cedex, France	(50)52-8102

Selecting the Oscilloscope

Table 8.3. Oscilloscope Manufacturers (*con't.*)

Tektronix Inc, Dept G, PO Box 500 Beaverton OR 97077	(503)644-0161
Tucker Elecs Co, Dept G, PO Box 401060 Garland TX 75040	(214)348-8800
Vu-Data Corp, 7170G Convoy Court San Diego CA 92111	(714)279-6572

OSCILLOSCOPES, PROGRAMMABLE

Continental Resources Inc, 175G Middlesex Turnpike Bedford MA 01730	(617)275-0850
Dumont Oscilloscope Labs Inc, 40G Fairfield Pl W Caldwell NJ 07006	(201)575-8666
E-H Intl Inc, 515G 11th St PO Box 1289 Oakland CA 94604	(415)834-3030
Norland Corp, Dept G, Rte 4 Norland Dr Ft Atkinson WI 53538	(414)563-8456

OSCILLOSCOPES, SAMPLING

Continental Resources Inc, 175G Middlesex Turnpike Bedford MA 01730	(617)275-0850
E-H Intl Inc, 515G 11th St PO Box 1289 Oakland CA 94604	(415)834-3030
Gould Inc, Instru Div Oscilloscope Prods, 3631G Perkins Av Cleveland OH 44114	(216)361-3316
Hewlett-Packard, Corporate Div, 1501G Page Mill Rd Palo Alto CA 94304	(415)856-1501
Philips Test & Measuring Instrus, 85G McKee Dr Mahwah NJ 07430	(201)529-3800
Tektronix Inc, Dept G, PO Box 500 Beaverton OR 97077	(503)644-0161
Wawasee Elecs, Dept G, PO Box 36 Syracuse IN 46567	(219)457-3191

OSCILLOSCOPES, STORAGE

Continental Resources Inc, 175G Middlesex Turnpike Bedford MA 01730	(617)275-0850
Electroplan Ltd, Dept G, PO Box 19 Orchard Rd, Royston, Herts SG8 5HH, England	Royston (0763)41171
Electro Rent Corp, 4131G Vanowen Pl Burbank CA 91505	(800)232-2173
General Atronics Corp, Dept G, 1200 E Mermaid Ln Phila PA 19118	(215)248-3700
General Electric Co, Quick Rental Instrus, 1G River Rd-Bldg 6-328 Schenectady NY 12345	(518)385-9950
Gould Advance Ltd, Instrus, Dept G, Roebuck Rd, Hainault, Essex 1G6 3UE, England	01-500-1000
Gould Inc, Instru Div Oscilloscope Prods, 3631G Perkins Av Cleveland OH 44114	(216)361-3316
Hameg, Dept G, 5-9 Av de la Republic, 94800 Villejuif, France	678.09.98
Hewlett-Packard, Corporate Div, 1501G Page Mill Rd Palo Alto CA 94304	(415)856-1501
Livingston Hire Ltd, Hire Elec Instru, 27-29G Camden Rd Shirley House, London NW1 9NR, England	01 2673262
Nicolet Instru Corp, 5225G Verona Rd Madison WI 53711	(608)271-3333
Norland Corp, Dept G, Rte 4 Norland Dr Ft Atkinson WI 53538	(414)563-8456
Philips Test & Measuring Instrus, 85G McKee Dr Mahwah NJ 07430	(201)529-3800
Rental Elecs Inc, 2445G Faber Pl Palo Alto CA 94303	(415)324-8080
Tektronix Inc, Dept G, PO Box 500 Beaverton OR 97077	(503)644-0161
Tucker Elecs Co, Dept G, PO Box 401060 Garland TX 75040	(214)348-8800

OSCILLOSCOPES, THREE DIMENSIONAL

Tektronix Inc, Dept G, PO Box 500 Beaverton OR 97077	(503)644-0161

OSCILLOSCOPOES, VECTORSCOPES

Continental Resources Inc, 175G Middlesex Turnpike Bedford MA 01730	(617)275-0850

Table 8.3. Oscilloscope Manufacturers (*con't.*)

Crow of Reading Ltd, Broadcast Div, Dept G, PO Box 36, Reading RG1 2NB, United Kingdom	UK + 734-595 025
Dynascan Corp, B & K Precision, Dept G, 6460 W Cortland Chicago IL 60635	(312)889-9087
Hickok Electrical Instru Co, Indl Instru, 10514G Dupont St Cleveland OH 44108	(216)541-8060
Infodex Inc, 7G Cherry Av Waterbury CT 06702	(203)757-9291
Leader Instrus Corp, 151G Dupont St Plainview NY 11803	(516)822-9300
Lectrotech Inc, Dept G, 5810 N Western Av Chicago IL 60659	(312)769-6262
Livingston Hire Ltd, Hire Elec Instru, 27–29G Camden Rd Shirley House, London NW1 9NR, England	01 2673262
Sencore Inc, 3200G Sencore Dr Sioux Falls SD 57107	(605)339-0100
Simpson Electric Co, Katy Inds, 853G Dundee Av Elgin IL 60120	(312)697-2260
Tektronix Inc, Dept G, PO Box 500 Beaverton OR 97077	(503)644-0161

OSCILLOSCOPE PROBES

American Laser Sys Inc, 106G James Fowler Rd Goleta CA 93017	(805)967-0423
Ballantine Labs Inc, Dept G, PO Box 97 Boonton NJ 07005	(201)335-0900
Continental Resources Inc, 175G Middlesex Turnpike Bedford MA 01730	(617)275-0850
Dumont Oscilloscope Labs Inc, 40G Fairfield Pl W Caldwell NJ 07006	(201)575-8666
Dynascan Corp, B & K Precision, Dept G, 6460 W Cortland Chicago IL 60635	(312)889-9087
Electroplan Ltd, Dept G, PO Box 19 Orchard Rd, Royston, Herts SG8 5HH, England	Royston(0763)41171
Electro Rent Corp 4131G Vanowen Pl Burbank CA 91505	(800)232-2173
Gould Advance Ltd, Instrus, Dept G, Roebuck Rd, Hainault, Essex 1G6 3UE, England	01-500-1000
Gould Inc, Instru Div Oscilloscope Prods, 3631G Perkins Av Cleveland OH 44114	(216)361-3316
Heath Co, Dept G, Benton Harbor MI 49022	(616)982-3200
Hewlett-Packard, Corporate Div, 1501G Page Mill Rd Palo Alto CA 94304	(415)856-1501
Leader Instrus Corp, 151G Dupont St Plainview NY 11803	(516)822-9300
Lectrotech Inc, Dept G, 5810 N Western Av Chicago IL 60659	(312)769-6262
Livingston Hire Ltd, Hire Elec Instru, 27–29G Camden Rd Shirley House, London NW1 9NR, England	01 2673262
Philips Test & Measuring Instrus, 85G McKee Dr Mahwah NJ 07430	(201)529-3800
Scopex Instrus Ltd, Dept G, Pixmore Indl Est, Letchworth Herts, England	04626-72771
Sencore Inc, 3200G Sencore Dr Sioux Falls SD 57107	(605)339-0100
Simpson Electric Co, Katy Inds, 853G Dundee Av Elgin IL 60120	(312)697-2260
SMK Elec Corp of America, Dept G, 118 E Savarona Way Carson CA 90746	(213)770-8915
Techni-Tool Inc, 5G Apollo Rd Plymouth Meeting PA 19462	(215)825-4990
Tektronix Inc, Dept G, PO Box 500 Beaverton OR 97077	(503)644-0161
Test Probes, Inc, Dept G, PO Box 2113 La Jolla CA 92038	
T & M Research Prods Inc, 129G Rhode Island St NE Albuquerque NM 87108	(505)268-0316
VIZ Mfg Co, Dept G, 335 E Price St Phila PA 19144	(215)844-2626
Vu-Data Corp, 7170G Convoy Court San Diego CA 92111	(714)279-6572

Selecting the Oscilloscope

Table 8.4. Instrument Rental Firms

Firm	Phone
Continental Resources Inc, 175 Middlesex Tpke Bedford, MA 01730	(617)275-0850
Electro Rent Corp, 4131 Vanowen Burbank, CA 91505	(213)843-3131
General Electric Co, Apparatus Service Div, 1 River Rd Bldg 4, Rm 210 Schenectady, NY 12345	(518)372-9900
Leasametric, 1164 Triton Dr Foster City, CA 94404	(415)574-4441
Rental Electronics Inc, 19525 Business Centre Dr Northridge, CA 91324	(415)856-7600
US Instruments Rental Inc 2121 S El Camino Real San Mateo, CA 94403	(415)574-6006

index

Ac voltage calibrator, 59
Accessories for oscilloscope, 54-68
AF equipment, tests and measurements, 107-144
AF signal tracer, use of oscilloscope as, 117
Alignment
  AM IF amplifier, 134-135
  FM detector, 136
  FM IF amplifier, 136
  TV front end, 138
  TV IF amplifier, 136-137
  TV sound IF amplifier and detector, 137
Alternate mode (dual-trace), 22-25
Alternating current, measurement of, 79-80
AM IF amplifier, visual alignment of, 134-135
Amplifier
  crosstalk, 154-155
  external, 61
  function, 10-11
  horizontal, 10, 18, 20-21, 155
  hum and noise level checking, 109-110
  phase shift, 111, 124-127, 154-155
  sync, 18, 37
  tests and measurements, 107-144
  vertical, 10, 18, 20-21, 25, 155, 161-162
  Z-axis, 20
Amplifier selector, 34
Amplitude modulation checking
  sine-wave method, 139-140
  trapezoidal-pattern method, 140-143
Astigmatism adjustment, 157-158
Astigmatism control, 41
Attenuator, 10, 36
  testing, 152-153

Bandwidth needs, 163-164
Basic layout of oscilloscope, 18-20
Blanking, 17
Bridge null detector, 117-119
Broken-line pattern, 92-94
Broken-ring pattern, 91-92

Calibrate, how to, 69-73
Calibrated time measurements, 94-95
Calibrated voltage measurements, 49-51
Calibrating voltage, 18, 20, 58
  deflection, 9
Calibration hints, 148
Calibration voltage control, 40-41
Calibrator
  ac voltage, 59
  dc voltage, 58-59
  frequency (time), 59
Camera, for trace photography, 62-67
Capacitor-type voltage-divider probe, 56
Cathode-ray tube. *See* CRT
Centering control
  horizontal, 34
  vertical, 34
Checking frequency-multiplier operation, 142-144
Checking sweep linearity, 156-157
Checking TV operating waveforms, 138-139
Checking video amplifier, 137-138
Chop mode (dual-trace), 22, 24-25
Coarse frequency control, 36
Collector voltage, 77
Color-bar signal, 29, 31-32
Color TV signal on vectorscope, 29
Comparing waveforms on dual-trace scope, 81-84, 98

Index

Compensation of attenuator, 152-153
Composite current, measurement of, 80
Composite voltage, measurement of, 77
Computer analysis and troubleshooting, 165
Control
  astigmatism, 41
  calibration voltage, 40-41
  coarse frequency, 36
  dc balance, 41, 151
  fine frequency, 37
  focus, 18, 34
  frequency compensation, 41
  horizontal centering, 34
  horizontal gain, 36
  hum balance, 41
  intensity, 18, 34
  level, 37
  linearity, 41
  phasing, 40
  slope, 37-38
  sweep delay, 36
  sweep frequency, 41
  sync amplitude, 37
  vertical centering, 34
  vertical gain, 36
  voltage regulation, 41
  Z-axis gain, 40
Controls and adjustments, 33-53
  functional, 33-41, 44-47
  operating, 33, 41
Crosstalk, 154-155
CRT, 2, 162-163
  complex tubes, 9
  deflecting plates, 3-4
  deflection voltages, 8-9
  features, 7-9
  operating voltages, 8
  phosphor, 8
  storage, 25
Current measurement
  ac, 79-80
  composite, 80
  dc, 79-80
  fluctuating, 80
Current probe, 80-81

Dc balance control, 41, 151
Dc voltage calibrator, 58-59
Deflection
  electromagnetic, 3
  electrostatic, 3
  sensitivity, 165-166
Deflection plates, 3-4
  direct access, 10
Delay line, 38-39, 122-123
Delayed sweep, 36, 159

Delayed time base, 39-41
Demodulator probe, 55-57
Desired-to-undesired signals, ratio of, 124
Differential voltage measurement, 84-87
Digital circuit checking, 98-101
Discharging electrified disk, 68
Distortion checking, 111-114, 119-121
Divide-by-eight circuit, 97
Divide-by-two circuit, 95-97
Dot-wheel pattern, 91-92
Driven sweep, 13, 36
Dual pattern, in phase measurement, 103-105
Dual-alternate display mode, 22-25
Dual-chopped display mode, 22, 24-25
Dual-trace oscilloscope, 21-25, 163, 165, 167-168
  checking phase angle between current and voltage, 105
  checking phase angle between two currents, 105-106
  digital circuit testing, 98-101
  frequency divider analysis with, 95-97
  setup procedure, 43-47

Electromagnetic deflection, 3
Electron beam, 2-3
  how it is deflected, 3-7
Electron gun, 2-3, 22, 25-26
Electronic switch, 60-61
Electrostatic deflection, 3
Erase control, 26
External amplifiers, 61

Film, for trace photography, 62-63, 65
Fine frequency control, 37
Flashlight analogy, 2
Flood gun, 25-26
Fluctuating current, measurement of, 80
Fluctuating voltage, measurement of, 77
Fluorescence, 2
Flyback, 17
FM detector, 136
FM IF amplifier alignment, 136
Focus control, 18, 34
Frequency calibrator, 59
Frequency compensation control, 41
Frequency divider analysis, 95-97
Frequency measurement and comparison, 88-106
  broken-line pattern, 92-94
  broken-ring pattern, 91-92
  Lissajous figures, 88-90

182 Index

Frequency measurement and comparison (*continued*)
　modulated-ring pattern, 90-91
　sawtooth internal sweep, 94
Frequency response, checking, 109
Frequency vernier, 37
Frequency-multiplier operation, checking, 142-144
Front-panel controls (functional controls), 33-41, 44-47
Fundamental suppressor, use of, in checking distortion, 112-114

Gain control
　horizontal, 36
　vertical, 36
　Z-axis, 40
Gated ringing circuit, 121
Gear-wheel pattern, 90-91
General-purpose oscilloscope, 8, 33-34
Generator distortion, checking, 111
Graticule, 9, 32
Grating, 9
Grid screen, 9

Hand tracing, from oscilloscope screen, 67-68
Harmonic distortion, checking, 111-114
High-voltage probe. *See* Voltage-divider probe
Horizontal amplifier, 20-21, 155
Horizontal centering control, 34
Horizontal deflection plates, 3-7
Horizontal gain control, 36
Hum balance control, 41
Hum level, checking, 109-110

IF amplifier alignment, 134-137
Input attenuator compensation, checking, 152-153
Input capacitance, checking, 152
Input resistance, checking, 152
Instruction manual, 145-146
Intensity control, 18, 34
Intensity modulation, 17-18, 20, 40
Intermodulation checking, 114-115
Internal adjustments of oscilloscope, 150-151

Layout of basic oscilloscope, 18-20
Layout of dual-trace scope, 22-25
Level control, 37
Linear sweep, 12
Linearity control, 41

Lissajous figures, 29, 31, 88-89, 101-103, 142-143
Locking a recurrent sweep, 14-17
Locking a triggered sweep, 14, 16
Low-capacitance probe, 55, 155

Magnifier, sweep, 38-39
Manufacturers of oscilloscopes, 173-178
Mask, 9
Mode switch, 38
Modulated-ring pattern, 90-91
Modulation
　amplitude tests, 139-143
　intensity, 17-18, 20, 40
　Z-axis, 17-18, 20
Modulator channel, checking, 143
Motion-picture camera, 67
Moving-film camera, 67
Multi-burst frequencies, 127
Multiple-channel oscilloscopes, 174
Multi-trace oscilloscopes, 174-175

Noise level, checking of, 109-110
Non-sawtooth sweep, 13

Operating controls, 33, 41
Operating precautions, 51-53
Oscilloscopes
　accessories, 54-68
　basic layout, 18-20
　basic steps in use of, 42
　controls and adjustments, 33-53
　conventional versus plug-in, 20-22, 161
　factors affecting buying decision, 22, 162
　leasing or renting, 169-170
　manufacturers, 173-178
　operating precautions, 51-53
　rental firms, 179
　selecting, 159-179
　servicing, 145-158

Persistence of trace, 8
Persistence of vision, 2
Phase angle between current and voltage, checking of, 105
Phase angle between two currents, checking of, 105-106
Phase measurement and comparison, 88-106
　angle between current and voltage, 105
　angle between two currents, 105-106

Index

Phase measurement and comparison (*continued*)
  dual pattern, 103-105
  Lissajous figures, 101-103
Phase shift of amplifier, checking, 111, 124-127, 154-155
Phase shift of oscilloscope, checking, 103
Phasing control, 40
Phosphors, 8, 25, 62
Photographing traces, 61-68
Plate deflection, 3-7
Plate voltage, 77
Plug-in oscilloscope, 20-22, 161, 173
Plug-in sections, 20-22
Polaroid Land camera, 63
Portable oscilloscope, 8, 175-177
Potentiometer, 36
Power output of amplifier, measuring, 110-111
Power supply ripple, 77-79
Precautions, operating, 51-53
Probes, 54-58
  capacitor-type voltage-divider, 56
  current probe, 80-81
  demodulator, 56-57
  low-capacitance, 55, 155
  manufacturers, 178-179
  resistor-type voltage-divider, 55-56
  RF, 57-58
Programmable oscilloscopes, 177
Propagation time measurement, 97-98
Pulsating voltage, measurement of, 76-77

Receiver, audio, 107-144
Receiver, color TV, 29-31
Recurrent sweep, 13-17, 166-167, 171
Resistor-type voltage-divider probe, 55-56
Retrace blanking, 17
RF probe, 55, 57-58
Risetime, 10-11

Sampling oscilloscope, 27-29, 177
Sawtooth sweep, 12-13, 94
Screen illumination, 34
Screens, 2, 9
  voltage calibration, 69-73
Setup procedures
  dual-trace scope, 43-47
  single-trace scope, 41-43
Signal tracer, use of oscilloscope as, 117
Sine-wave method of amplitude modulation checking, 139-140
Single sweep, 13, 36

Single-trace scope, 167
  setup procedures, 41-43
Single-trace waveform, observation on dual-trace scope, 47-49
Slope control, 37-38
Source switch, 38
Spot-wheel pattern, 91-92
Square-wave testing, 10-11, 115-117, 137-138, 153-155
Stereo servicing, 123
Storage oscilloscopes, 25-27, 160, 177
Sweep
  delayed, 36, 159
  driven, 13, 36
  linear, 12
  linearity check, 156-157
  magnification, 38-39
  non-sawtooth, 13
  range selector, 36
  recurrent, 13-17
  sawtooth, 12-13
  selector, 36
  single, 13, 36
  synchronization, 17
  triggered, 13-17
  V and H settings, 156
Sweep frequency, 14-15
Sweep frequency control, 37, 41
Sweep frequency drift, 17
Sweep generator
  function of, 11-17
  synchronization, 17
Sweep magnifier, 38-39
Sweep range, 164-165
Sweep range selector, 36
Sweep selector, 36
Sweep speed, 25
Sweep voltage, 12-14
Sync amplifier, 37
Sync amplitude control, 37
Sync selector, 37
Synchronization, 17, 37

Three dimensional oscilloscope, 177
Time base, 166-167
Time calibrator, 59
Toothed-wheel pattern, 90-91
Trace photography, 61-68
Trace reverser, 34
Transmitter, tests and measurements, 107-144
Transmitter modulation checking, 18
Trapezoidal-pattern method of amplitude modulation checking, 140-143
Triggered scope, 24, 37-38, 166, 172
Triggered sweep, 13-17

184   *Index*

Troubleshooting chart, 148-150
Troubleshooting the scope, 147-148
Tunable distortion meter, 114
TV IF amplifier alignment, 136-137
TV front end alignment, 138
TV operating waveforms, checking, 138-139
TV sound IF amplifier and detector alignment, 137
TV video amplifier, checking, 137-138

Variable-delay time bases, 161
Vectorgram, 29
Vectorscope, 29-32, 177-178
Vertical amplifier, 20-21, 25, 155, 161-162
Vertical blanking interval, 127-128
Vertical centering control, 34
Vertical deflection plates, 3-7
Vertical gain control, 36
Vertical interval test signal. *See* VITS
Video amplifier, checking, 137-138
Viewing screens, 9
VITS signal, 126-134
Voltage
  amplitude, 11-12
  calibrating, 9, 18, 20, 58
  direct measurement of, 73-74
  pulsating, measurement of, 76-77
  sweep, 11-17
Voltage calibration of screen, 69-73
  ac, 69-73
  dc, 72-73
Voltage calibrator
  ac, 59
  dc, 58-59

Voltage divider probe, 55
  capacitor-type, 56
  resistor-type, 55-56
Voltage gain or loss, checking, 108-109
Voltage measurement
  differential, using dual-trace scope, 84-87
  with voltage calibrator, 75-76
Voltage regulation control, 41

Waveform
  comparisons, on dual-trace scope, 81-84
  divide-by-eight circuit, 97
  time-related, 98-101
Waveform photography. *See* Trace photography
Wave shape, checking, 107
Wideband oscilloscope, 33
Write gun, 25-26
Writing rate, in oscilloscope photography, 64

X-gain control, 36
X-position control, 34

Y-delay line, 122-123
Y-gain control, 36
Y-position control, 34

Z-axis amplifier, 20
Z-axis gain control, 40
Z-axis modulation, 17-18, 20
Zener-diode controlled calibration, 156-157